# Springer Tracts in Modern Physics

Springer Tracts in Modern Physics provides comprehensive and critical reviews of topics of current interest in physics. The following fields are emphasized: elementary particle physics, solid-state physics, complex systems, and fundamental astrophysics.
Suitable reviews of other fields can also be accepted. The editors encourage prospective authors to correspond with them in advance of submitting an article. For reviews of topics belonging to the above mentioned fields, they should address the responsible editor, otherwise the managing editor.
See also http://www.springer.de/phys/books/stmp.html

# Springer Tracts in Modern Physics
# Volume 179

Now also Available Online

Starting with Volume 165, Springer Tracts in Modern Physics is part of the Springer LINK service. For all customers with standing orders for Springer Tracts in Modern Physics we offer the full text in electronic form via LINK free of charge. Please contact your librarian who can receive a password for free access to the full articles by registration at:

http://link.springer.de/series/stmp/reg_form.htm

If you do not have a standing order you can nevertheless browse through the table of contents of the volumes and the abstracts of each article at:

http://link.springer.de/series/stmp/

There you will also find more information about the series.

Springer
*Berlin*
*Heidelberg*
*New York*
*Hong Kong*
*London*
*Milan*
*Paris*
*Tokyo*

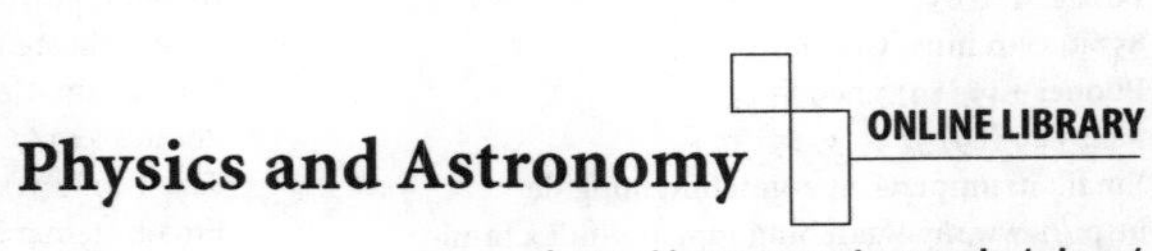


http://www.springer.de/phys/

Yoshio Waseda

# Anomalous X-Ray Scattering for Materials Characterization

## Atomic-Scale Structure Determination

With 132 Figures

Springer

Professor Yoshio Waseda
Institute for Advanced Materials Processing
Tohoku University
Katahira-2-choume 1-1, Aoba-ku, Sendai 980-8577, Japan
E-mail: waseda@iamp.tohoku.ac.jp

Library of Congress Cataloging-in-Publication Data applied for.

Die Deutsche Bibliothek - CIP-Einheitsaufnahme
Waseda, Yoshio:
Anomalous X-ray scattering for materials characterization : atomic scale structure determination / Yoshio Waseda. - Berlin ; Heidelberg ; New York ; Hong Kong ; London ; Milan ; Paris ; Tokyo : Springer, 2002 (Springer tracts in modern physics ; Vol. 179)
(Physics and astronomy online library)
ISBN 3-540-43443-7

Physics and Astronomy Classification Scheme (PACS): 61.10, 61.25, 07.85, 81.00

ISSN print edition: 0081-3869
ISSN electronic edition: 1615-0430
ISBN 3-540-43443-7 Springer-Verlag Berlin Heidelberg New York



Springer-Verlag Berlin Heidelberg New York
a member of BertelsmannSpringer Science+Business Media GmbH

http://www.springer.de



Typesetting: LaTeX-data from the author
Production: LE-TeXJelonek, Schmidt & Vöckler GbR, Leipzig
Cover design: *design & production* GmbH, Heidelberg

Printed on acid-free paper SPIN: 10786894 57/3141/YL 5 4 3 2 1 0

To
Yuko Waseda
with devotion and dedication

# Preface

The production of multi-layered thin films with sufficient reliability is a key technology for device fabrication in micro-electronics. In the Co/Cu-type multi-layers, for example, magnetoresistance has been found as large as 80 % at 4.2 K and 50 % at room temperature. In addition to such gigantic magnetoresistance, these multi-layers indicate anti-ferromagnetic and ferromagnetic oscillation behavior with an increase in the thickness of the layers of the non-magnetic component. These interesting properties of the new synthetic functional materials are attributed to their periodic and interfacial structures at a microscopic level, although the origin of such peculiar features is not fully understood. Information on the surface structure or the number density of atoms in the near-surface region may provide better insight.

Amorphous alloys, frequently referred to as metallic glasses, are produced by rapid quenching from the melt. The second-generation amorphous alloys, called "bulk amorphous alloys", have been discovered in some Pd-based and Zr-based alloy systems, with a super-cooled liquid region at more than 120 K. In these alloy systems, one can obtain a sample thickness of several centimeters. Growing scientific and technological curiosity about the new amorphous alloys has focused on the fundamental factors, such as the atomic-scale structure, which are responsible for the thermal stability with certain chemical compositions. However, it is not easy to document the lack of crystal-like atomic periodicity that is probably responsible for the large metastable region. Since the bulk amorphous alloys usually contain several types of atoms (at least three components are always present), only a weighted sum of pair correlations (partial structures) of the individual chemical constituents can be obtained from conventional X-ray diffraction. The nearest neighbor atomic correlations of the individual chemical constituents or the local environmental structure around a specific element is required for describing the quantitative structure in multi-component, non-crystalline systems of interest.

Since the importance of the atomic-scale structure for advanced materials has been well recognized from both basic and applied science points of view, several techniques using X-rays and neutrons have been employed in the determination of the fine structure. Each method has its own advantages and disadvantages. Nevertheless, the anomalous X-ray scattering (AXS) method, utilizing the so-called anomalous dispersion effect near the absorption edge of the constituent element, is considered to be the most powerful method for obtaining the accurate partial structure functions for individual pairs of chemical constituents or the environmental structure functions around a specific element in multi-component systems. Current systematic AXS studies using synchrotron radiation as the source show great promise. The AXS method is useful for both crystalline and non-crystalline systems for the analysis of bulk and surface structures. The AXS method reveals the structural features of individual chemical constituents even for systems containing neighboring elements in the periodic table, such as the distribution of cobalt in complex ferrites or in amorphous ferrite glasses. However, no specialized monograph on this relatively new method for structural characterization is available except for a few conference proceedings.

This book first presents the fundamentals of the anomalous dispersion effect in the close vicinity of the absorption edge of K- or L-shell electrons of a constituent element. Next, the experimental apparatus suitable for structural studies applying this relatively new method is explained. Then, the validation of the AXS method and its application to a variety of different materials is demonstrated. The examples include molten salts, liquid alloys, solutions, amorphous alloys, oxide glasses, super-ionic conducting glasses, amorphous thin films, quasi-crystals and ultra-fine powders. The results obtained by using the anomalous small-angle X-ray scattering (SAXS) or anomalous grazing-incidence X-ray reflection (GIXR) method are also provided with some selected examples of the GP zone, composition modulation, the specific volume in bulk materials or multi-layered thin films. The relative merits of the method and its future prospects are also discussed. This is the first major compilation of current advances in anomalous X-ray scattering as a major new tool for the structural characterization of materials in a variety of states. The anomalous dispersion factors including mass-absorption coefficients for 96 elements in the energy range from 1 to 50 keV are given in our public database, SCM-AXS (http://www.iamp.tohoku.ac.jp/; http://www.tagen.tohoku.ac.jp/index-j.html). Such information is always required for carrying out successful anomalous X-ray scattering on crystalline and non-crystalline materials. This information is not available in the previous specialized monographs and common data-books for X-ray crystallography. The author, therefore, believes that the present book, with many references, provides an adequate guide for both specialists and non-specialists who wish to become acquainted with the new tool for the structural characterization of materials.

Many people have helped, directly or indirectly, in preparing this book. The author is deeply indebted to Professors S. Tamaki, T. Egami and the late S. Hosoya for valuable discussions regarding various aspects of AXS and non-crystalline systems such as liquids and amorphous solids. I am also grateful to Professors H. Iwasaki and T. Masumoto for their sustained encouragement for our serial research projects on the structure and properties of various materials at the Institute of Multidisciplinary Research for Advanced Materials (formerly the Institute for Advanced Materials Processing), Tohoku University, Japan. A significant part of the AXS results compiled in this monograph is based on joint research involving the author and Professors E. Matsubara (Tohoku University), K. Sugiyama (Tokyo University) and M. Saito (Niigata University) and the former Ph.D. students Drs. K. Shinoda, A.H. Shinohara, T. Kosaka, S. Sato, C.Y. Park and S.C. Kang. Their contributions as well as the dedicated service and advice given by the staff members at the Photon Factory of the Institute of Materials Structure Science in the High Energy Accelerator Research Organization (Tsukuba/Japan), particularly Professors T. Matsushita and M. Nomura, are gratefully acknowledged. The author also would like to mention that these many results could not have been obtained without the prior experience gained using the synchrotron radiation as a light source at the Cornell High Energy Synchrotron Source (CHESS), Cornell University (Ithaca/USA). So the assistance given by Professor B.W. Batterman and Drs. D. Bilderback and D. Mills in collaborative research with Professor T. Egami (University of Pennsylvania) and Drs. S. Aur, D. Kofalt (University of Pennsylvania), H.S. Chen and B.K. Teo (Bell Laboratories) is gratefully acknowledged. Professor H. Ohta (Ibaraki University) assisted me in the calculation of the energy dependence of the anomalous dispersion factors for 96 elements. Many thanks are due to Professor K.T. Jacob (Indian Institute of Science) and Professor N.J. Themelis (Columbia University), who read parts of the manuscript and provided many helpful suggestions. I would also like to give my thanks to Dr. H. Shibata, Ms. N. Eguchi and Mr. Y. Ito for their assistance in preparing the electronic typeset version of this monograph. The author is also indebted to the many sources of material cited in this book. Finally, this book is dedicated to my wife, Yuko Waseda, although this is indeed very little compensation for her many sacrifices and patience during the extended period of research for and composition of this book.

Sendai, Japan
April 2002

*Yoshio Waseda*

# Contents

# 1. Structural Characterization of Crystalline and Non-crystalline Materials – A Brief Background of Current Requirements

The X-ray powder diffraction technique is a well-established and widely used method for both qualitative and quantitative analysis of various substances in a variety of states (see, for example, [1]). However, in a multi-component mixture with a relatively complicated chemical composition, we frequently find difficulty in identifying the individual chemical constituents by the conventional X-ray powder diffraction method. There are also generally insufficient differences in the X-ray diffraction intensities for two elements of nearly the same atomic number in the periodic table. For example, this is certainly the case for a mixture of copper sulfide and ferrite components in the products of a copper smelting process.

Ferrites are a group of compounds with "spinel" structure [2] expressed by the general formula $MFe_2^{3+}O_4$, where M is a divalent cation. They are known to have very interesting electrical and magnetic properties which are controlled by the distribution of cations between different sites. Substituting one element for another is very common in materials processing for controlling new functional properties of specific compounds. A clear understanding of the physical and chemical properties of such oxide materials depends heavily on their atomic-scale structure. In such ferrite materials, described by the "spinel" structure, there are 32 octahedral and 64 tetrahedral interstices formed by oxygen atoms available for cations in a unit cell, and half of the octahedral sites and one-eighth of the tetrahedral sites are known to be occupied. For example, a divalent zinc cation ($Zn^{2+}$) usually prefers the tetrahedral sites in zinc ferrite ($ZnFe_2O_4$) and is of the *normal* type [3]. On the other hand, magnetite ($Fe_3O_4$) is classified as *inverse* spinel, where the tetrahedral sites contain only ferric ion ($Fe^{3+}$); the residual $Fe^{3+}$ and ferrous ($Fe^{2+}$) ions are octahedrally coordinated at low temperatures [4]. Since the magnetic properties of ferrite spinels are very sensitive to the cation distribution, it is of great importance to determine their degree of inversion. However, determination of the cation distribution in these ferrite materials is not easy, because the X-ray scattering abilities of the components M, such as Zn and Ni, are close to that of the host element, Fe.

The use of the anomalous X-ray scattering (hereafter referred to as AXS) method at energies near the absorption edge of the M component [5] is undoubtedly one way to overcome the experimental difficulties, by making available sufficient atomic sensitivity arising from the so-called "anomalous dispersion effect" near the absorption edge or by providing an appreciable difference in the crystallographic structure factors [1, 5]. This applies even for two elements of close atomic numbers in the periodic table, such as Fe and Ni.

The advantage of the AXS method in the structural analysis of crystalline materials has been suggested in the past, but the AXS results were still limited to a small number of compositions. However, a synchrotron radiation source of X-rays recently became available for applying to materials characterization and has dramatically improved the quality of the AXS data relative to that obtained with a conventional X-ray source. Although "beam time" in a synchrotron facility is scarce, the AXS method is of great benefit to the analysis of various crystalline materials.

## 1.1 Site Occupancy

The determination of the site occupancy (or space group) in a complex system consisting of more than two elements is not an easy task, even when using Rietveld analysis [6], because convergence is often not obtained even after many iterations. In such a case, the AXS measurement is very useful, because the intensity variation detected at two energy levels in the close vicinity of the absorption edge of a specific element, M, in a sample should be attributed to the contribution originating only from M. The anomalous dispersion effects arising from other elements appear to be insignificant in the corresponding energy region.

In 1986, Bednorz and Müller [7] discovered superconductivity above 30 K in the Ba–La–Cu–O system. Their finding generated an enormous amount of activity in the materials science and engineering community. Many subsequent works indicated that several oxide systems such as Ba–Y–Cu–O [8] and Bi–Tl–Cu–O [9], have a superconducting transition temperature higher than the liquid nitrogen temperature (77 K). It would be very stimulating to extend these new oxide materials to practical applications for various devices. However, some reservations are frequently expressed regarding the quantitative accuracy of their fundamental structure, because of the many possibilities arising from the combination of more than four components plus defects. The AXS method holds promise in reducing this difficulty by making possible a comparison of the intensity variation between calculations based on the oxygen-deficient perovskite atomic arrangement and the measured intensity profile.

## 1.2 Quasi-crystals

The discovery by Shechtman et al. [10] in 1984 of aluminum-based alloys with icosahedral point-group symmetry and long-range orientational order stimulated many theoretical and experimental studies of this subject. Many other ternary alloys forming the icosahedral phase or decagonal phase in the two-dimensional case have been reported and classified into a relatively new category of "quasi-crystals". However, atoms in the quasi-crystals have no translational and rotational symmetries, so that an infinitely large unit cell is required to describe the atomic-scale structure. This makes structural analysis for quasi-crystals extremely complicated. One of the successful approaches, proposed by Henley [11], describes the atomic structure in a quasi-crystal by placing atoms on a rigid geometrical frame with a certain decoration rule. The AXS method has been found to be quite useful in studying the decoration rule in quasi-crystals.

## 1.3 Liquids and Glasses

The physics and chemistry of so-called non-crystalline (or disordered) materials, in which the atomic arrangement is not spatially periodic as in the case of crystalline materials, are also well recognized as an important and promising branch for materials research. Typical examples of non-crystalline materials are liquids and glasses of condensed matter. Current interest in this field arises mainly from the development of amorphous alloys (or metallic glasses) produced by rapid quenching from the melt (see, for example, [12]), because of their technological potential for application in soft magnetic elements, electronic devices and excellent high-tension wires with good corrosion resistance. The discovery of bulk amorphous alloys [13] of thickness on the order of several centimeters has provided great impetus in the study of this relatively new class of non-crystalline materials. Most advances have been made recently [14], although this research field itself has been studied for the last 30 years.

Liquid metals, salts and oxide mixtures (slags) are known to play a significant role in many metallurgical processes (see, for example, [15]). Some liquid alkali metals are potential heat-transfer media in nuclear-energy generation. Noble-metal halides such as CuBr and AgI are known to have a super-ionic conducting phase, indicating potential electrolytes [16]. Their growing technological importance and the novelty of the physics, mainly related to the non-periodicity in their atomic arrangement, have led to an increasing need for a better description of the atomic-scale structure and greater understanding of their various properties at a microscopic level.

## 1.4 Environmental Structure Around a Specific Element

Quantitative description of the atomic-scale structure of non-crystalline materials usually employs the *radial distribution function* (RDF), which indicates the probability of finding another atom at a distance from an origin atom as a function of the radial distance obtained by averaging spherically (see, for example, [17]). The information provided by the RDF is only one-dimensional, but it does describe quantitatively the atomic arrangements in non-crystalline materials. X-ray diffraction has been widely used to obtain the RDF of a variety of materials, but the structural studies for non-crystalline materials, except for one-component systems, are far from complete for several reasons. The environment of each atom in non-crystalline systems including more than two components generally differs from those of other atoms. This makes the interpretation of their RDFs difficult. Furthermore, the structure–property relationships of multi-component, non-crystalline systems can be determined only on the basis of the full set of the partial RDFs for the individual chemical constituents. In an A–B binary system, we need three partial RDFs for the A–A, B–B and A–B pairs and another six partials for the ternary case. Therefore, the utmost importance of the determination of partial structure functions is well recognized as one of the most essential research subjects for non-crystalline materials involving more than two components (see, for example, [18]). However, the actual implementation of this subject is not a trivial task even for a binary system.

Several methods for extracting the partial structure factors, corresponding to the Fourier transform of RDFs, have been proposed (see, for example, [19]), and a large amount of experimental and theoretical effort has been devoted to this research field. For example, the partial structure factors for a binary system can be estimated by making available at least three independent intensity measurements for which the weighting factors are varied without any change in their atomic distribution. The isotope substitution method for neutron diffraction, first applied by Enderby et al. in 1966 to liquid Cu–Sn alloys [20] and thereafter to several molten salts (see, for example, [21]), is considered to be one of the powerful methods. However, it is somewhat limited in practice by the lack of suitable isotopes; also the structure is automatically assumed to remain identical upon substitution of the isotopes. In this regard, use of the AXS method will, in the author's view, overcome this difficulty without requiring any assumptions and allow many more elements in the periodic table to be studied.

The AXS method can provide information about the local chemical environment of a specific element, which is of course quite important for quantitative determination of particular properties of non-crystalline materials at a microscopic level. Such environmental structure information obtained by the AXS method is very similar to the results of the so-called Extended X-ray Absorption Fine Structure (EXAFS, or simply called XAFS) measurement [22]. However, we are rather convinced that the AXS method is much more straightforward, at least theoretically, and provides environmental structure information including so-called middle-range ordering, as a function of radial distance, with much higher reliability than the EXAFS method. The EXAFS method is undoubtedly one of the most powerful methods for determining the local atomic structure in near neighbors of various materials. However, as noted by Lee et al. in 1981 [23], EXAFS does not differentiate easily between a reduction in the short-range-order parameter and the degree of disorder unless a considerable amount of fundamental structure information is already known about the desired materials. Therefore, it is *unrealistic* to expect *the EXAFS method alone* to provide the correct structure information for a completely unknown and complex material. For this reason, the AXS data could, at least, supplement the interpretation of the EXAFS data or *vice versa*. In fact, the AXS method may be a very reliable and powerful tool for determining the fine structure in multi-component, non-crystalline materials.

## 1.5 Small-Angle X-ray Scattering

Small-angle X-ray scattering [24] (hereafter referred to as SAXS) results enable us to establish many important microstructure parameters in a sample of interest, such as the particle volume, the nature of GP zones in Al-based alloys, the decomposition modulation in alloys, and the particle shape producing structural inhomogeneity. The determination of the partial structure functions in ternary alloys and the specific volume ratio in multi-layers is also very useful. SAXS studies are usually made using radiation at an energy level that is far from the absorption edge of any constituent element in a sample. Since the interpretation of the SAXS data, in principle, depends on the models used for theoretical calculation of the intensity, it is frequently found that more than two kinds of models can fit the experimental data equally well. The SAXS measurements coupled with AXS can overcome this obstacle, because significant improvement in changing the scattering contrast of a desired element can be obtained. This method was successfully used for characterizing the structure of materials and providing information that could not be obtained by the conventional SAXS method. This includes accurate determination of the periodic structure of multi-layered thin films [25].

## 1.6 Surface and Layered Structure

The production of multi-layered thin films with sufficient reliability is well recognized as a key technology for device fabrication in micro-electronics. With remarkable progress in such fields, X-ray optical methods such as grazing-incidence X-ray diffraction (GIXD) and grazing X-ray reflectometry (GXR) are widely used to investigate the structural properties of various multi-layered film materials. However, for use in structural characterization, often these techniques require the determination of the atomic number density of constituents or that of a near-surface component, especially with regard to unknown materials. Although X-ray photoelectron spectroscopy (XPS), Auger electron spectroscopy (AES) and secondary ion mass spectroscopy (SIMS) techniques are extensively applied for determining the composition of thin films with good sensitivity to the surface, they give only the relative quantities of constituents. There are also destructive probes for obtaining the compositional depth profiles in the multi-layered film sample by sputtering.

With respect to this particular subject, GXR with AXS, frequently referred to as the AGXR method, appears to be one way to determine the absolute value of the atomic number density in materials non-destructively [5,26]. The AGXR method is based on measuring the deviation in the refractive index of a substance of interest through the anomalous dispersion phenomena, and its usefulness was first demonstrated by a single-layered thin film grown on a glass substrate [27]. Recently, the capability of the AGXR method has been tested by obtaining the atomic number densities of constituents in a multi-layered thin film consisting of a GaAs/AlAs/GaAs heterostructure, when coupled with the Fourier filtering technique [28].

AXS is applicable to various crystalline and non-crystalline materials with only a few exceptions, such as the light elements. This advantage contrasts with other techniques, such as neutron diffraction using anomalous scattering or isotope substitution. The intense white X-ray source of synchrotron radiation produced from a multi-GeV electron storage ring is now available in many countries: USA, Germany, England, France, Italy, Japan, Korea, Brazil, Thailand and others. This situation has greatly improved both the acquisition and quality of the AXS data by enabling the use of an energy in which the AXS effect is maximized. Therefore, it may be suggested from considering many factors that the usefulness and potential power of the AXS method, in the author's view, cannot be overemphasized in answering various questions unsolved by the conventional X-ray diffraction method.

## References

1. B.D. Cullity: *Elements of X-ray Diffraction* (2nd edition) (Addison-Wesley, Reading 1978)
2. F.S. Galasso: *Structure and Properties of Inorganic Solids* (Pergamon, Oxford 1970)
3. H.St.C. O'Neil: Eur. J. Miner. **4**, 571 (1992)
4. M.E. Fleet: Acta Crystallogr., B **37**, 917 (1981)
5. R.W. James: *The Optical Principles of the Diffraction of X-rays* (G.Bells, London 1954)
6. H.M. Rietveld: J.Appl. Crystallogr., **2**, 65 (1969)
7. J.G. Bednorz and K.A.Müller: Z. Phys., **64**, 189 (1986)

8. H. Takagi, S. Uchida, K. Kishio, K. Kitazawa, K. Fueki and S. Tanaka: Jpn. J.Appl. Phys., **26**, L320 (1978)
9. J. Akimitsu, A. Yamazaki, H. Sawa and H. Fujiki: Jpn. J. Appl. Phys., **26**, L2080 (1987)
10. D. Shechtman, I.A. Blech, D. Gratias and J.W. Cahn: Phys. Rev. Lett., **53**, 1951 (1984)
11. C.L. Henley: Commun. Condens. Matter Phys., **13**, 59 (1987)
12. F.E. Luborsky: *Amorphous Metallic Alloys* (Butterworth, London 1983)
13. A. Inoue, N. Nishiyama and H. Kimura: Mater. Trans. JIM, **37**, 179 (1997)
14. A. Inoue: *Bulk Amorphous Alloys, Practical Characteristics and Applications* (Trans. Tech. Uetkon-Zurich 1999)
15. F.D. Richardson: *Physical Chemistry of Melts in Metallurgy* (Academic Press, London 1974)
16. J.B. Boyce and B.A. Huberman: Phys. Rep., **51**, 189 (1979)
17. T.L. Hill: *Statistical Mechanics* (McGraw-Hill, New York 1956)
18. C.N.J. Wagner: *Liquid Metals, Chemistry and Physics* (edited by.S.Z. Beer, Marcel-Dekker, New York 1972) pp. 258
19. D.T. Keating: J.Appl. Phys., **34**, 923 (1963)
20. J.E. Enderby, D.M. North and P.A. Egelstaff: Philos. Mag., **14**, 961 (1966)
21. D.I. Page and I. Mika: J. Phys. C., **4**, 3034 (1971)
22. B.K. Teo: *EXFAS Basic Principles and Data Analysis* (Springer, Berlin, Heidelberg, New York 1986)
23. P.A. Lee, P.H. Citrin, P. Eisenberger and B.M. Kincaid: Rev. Mod. Phys., **53**, 761 (1981)
24. A. Gunier: *Theory and Techniques for X-ray Crystallography* (Dumond, Paris 1964)
25. K. Kato, E. Matsubara, M. Saito, T. Kosaka, Y. Waseda and K. Inomata: Mater. Trans. JIM, **36**, 408 (1995)
26. W.C. Marra, P. Eisenberger and A.Y. Cho: J. Appl. Phys., **50**, 6972 (1979)
27. M. Saito, E. Matsubara and Y. Waseda: Mater. Trans. JIM, **37**, 39 (1996)
28. M. Saito and Y. Waseda: Mater. Trans. JIM, **40**, 1044 (1999)

# 2. Experimental Determination of Partial and Environmental Structure Functions in Non-crystalline Systems – Fundamental Aspects

All atomic positions in crystalline materials are described by means of a few parameters of distance and angle. However, such a simple definition is impossible in non-crystalline systems such as liquids and glasses, because of the lack of long-range structure periodicity. However, the atomic-scale structure of non-crystalline systems can be described quantitatively in terms of the so-called radial distribution function (RDF), which indicates the average probability of finding another atom within a specified volume at a distance from an origin atom as a function of the radial distance. The RDF information gives spherically averaged information on the atomic correlation as one-dimensional data; however, it provides unique quantitative information for describing the structure without long-range periodicity. In other words, the method is somewhat limited for describing the structure of non-crystalline systems. The description of the principles and the utility of the RDF has already been published in detail (see, for example, [1,2]); so here we introduce the essential points of the RDF analysis of non-crystalline systems for the convenience of discussion.

In the case of an hypothetical, homogeneous, non-crystalline system, the radial distribution function, $\mathrm{RDF} = 4\pi r^2 \rho(r)$, may be defined by considering a spherical shell of radius $r$ with thickness $\mathrm{d}r$ centered on an origin atom. The quantity $\rho(r)$ is often referred to as the radial density function corresponding to the average probability of finding another atom as a function of only distance. As shown in Fig. 2.1, the RDF gradually approaches the parabolic function $4\pi r^2 \rho_\circ$ at a larger value of $r$, where $\rho_\circ$ is the average number density of atoms, because the positional atomic correlation disappears with increasing distance in non-crystalline systems. It may be safely said that no atomic correlation exists within the minimum nearest-neighbor distance such as the atomic core diameter, arising from the repulsion in the pair potential. Therefore, the RDF should be equal to zero at such small values of $r$. The area under the respective peak in the RDF yields information about the coordination number on an average. It is worth mentioning that the concept of the RDF is applicable to any crystalline system where the atoms occupy the cube corners of a regular three-dimensional lattice. Of course, in such systems, the RDF is characterized by the discrete sharp peaks with fixed coordination

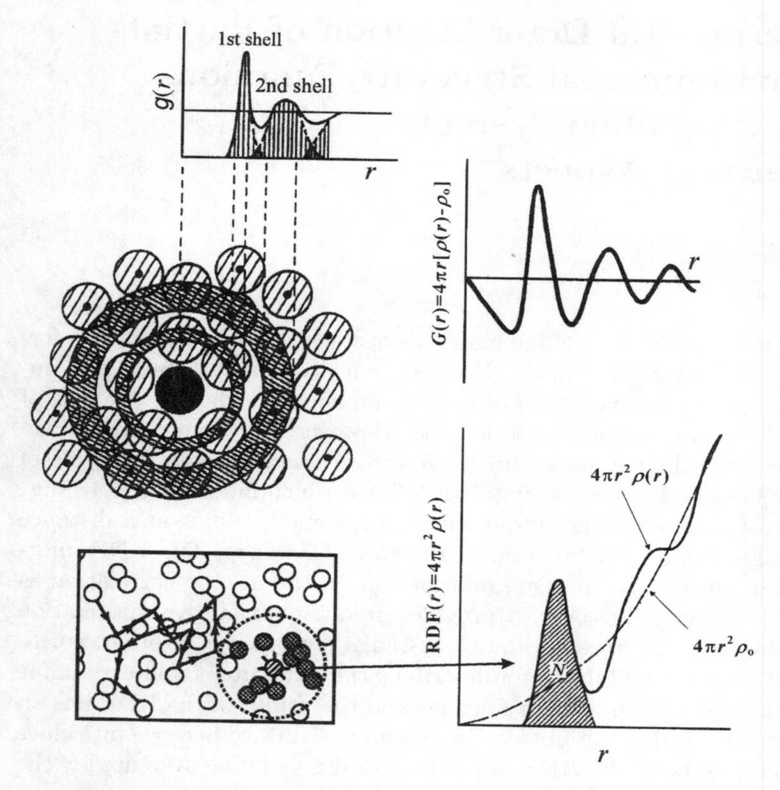


**Fig. 2.1.** Schematic of a snapshot of the atomic distribution and its RDF in a non-crystalline system

numbers, as shown in Table 2.1 using five simple crystal structures as an example.

The reduced RDF of $G(r)$ expressed by the following equation is also used widely for describing the atomic-scale structure of non-crystalline systems (see, for example, [3]):

$$G(r) = 4\pi r[\rho(r) - \rho_\circ]. \tag{2.1}$$

The function $g(r) = \rho(r)/\rho_\circ$, referred to as the pair distribution function, is also frequently used. It may be noted that $g(r)$ is sometimes also called the RDF.

Even in non-crystalline systems with a lack of long-range periodicity, the scattered beams from two atoms coherently interfere with each other and the scattering intensity depends on the relative positions of the two atoms. For this reason, the RDF in a non-crystalline system can be determined from diffraction data with X-rays, neutrons and electrons. The X-ray case is discussed below as an example, although many of the concepts and procedures

**Table 2.1.** Atomic distances and their coordination numbers, $N_j$, in some crystalline systems. $r_1$: nearest-neighbor distance; $a, c$: lattice constants

| | fcc | | hcp | | bcc | | Cubic | | Diamond | |
|---|---|---|---|---|---|---|---|---|---|---|
| | $r_1 = \frac{\sqrt{2}}{2}a$ | | $r_1 = \sqrt{\left(\frac{a^2}{3} + \frac{c^2}{4}\right)}$ | | $r_1 = \frac{\sqrt{3}}{2}a$ | | $r_1 = a$ | | $r_1 = \frac{\sqrt{3}}{4}a$ | |
| $j$ | $N_j$ | $(r_j/r_1)^2$ | $N_j$ | $(r_j/r_1)^2$ | $N_j$ | $(r_j/r_1)^2$ | $N_j$ | $(r_j/r_1)^2$ | $N_j$ | $(r_j/r_1)^2$ |
| 1 | 12 | 1 | 12 | 1 | 8 | 1 | 6 | 1 | 4 | 1 |
| 2 | 6 | 2 | 6 | 2 | 6 | $1\frac{1}{3}$ | 12 | 2 | 12 | $2\frac{2}{3}$ |
| 3 | 24 | 3 | 2 | $2\frac{2}{3}$ | 12 | $2\frac{2}{3}$ | 8 | 3 | 12 | $3\frac{2}{3}$ |
| 4 | 12 | 4 | 18 | 3 | 24 | $3\frac{2}{3}$ | 6 | 4 | 6 | $5\frac{1}{3}$ |
| 5 | 24 | 5 | 12 | $3\frac{2}{3}$ | 8 | 4 | 24 | 5 | 12 | $6\frac{1}{3}$ |
| 6 | 8 | 6 | 6 | 4 | 6 | $5\frac{1}{3}$ | 24 | 6 | 24 | 8 |
| 7 | 48 | 7 | 12 | 5 | 24 | $6\frac{1}{3}$ | 12 | 8 | 16 | 9 |
| 8 | 6 | 8 | 12 | $5\frac{2}{3}$ | 24 | $6\frac{2}{3}$ | 30 | 9 | 12 | $10\frac{2}{3}$ |
| 9 | 36 | 9 | 6 | 6 | 24 | 8 | 24 | 10 | 24 | $11\frac{2}{3}$ |
| 10 | 24 | 10 | 6 | $6\frac{1}{3}$ | 32 | 9 | 24 | 11 | 24 | $13\frac{1}{3}$ |
| 11 | 24 | 11 | 12 | $6\frac{2}{3}$ | 12 | $10\frac{2}{3}$ | 8 | 12 | 12 | $14\frac{1}{3}$ |
| 12 | 24 | 12 | 24 | 7 | 48 | $11\frac{2}{3}$ | 24 | 13 | 8 | 16 |
| 13 | 72 | 13 | 6 | $7\frac{1}{3}$ | 30 | 12 | 48 | 14 | 24 | 17 |

are similarly applicable to the measurements found using neutron and electron diffraction.

The diffraction wave vector, $Q$, is expressed in the following form:

$$Q = 4\pi \sin\theta/\lambda \tag{2.2}$$

$$= \left(4\pi/hc^\circ\right) \sin\theta \cdot E, \tag{2.3}$$

where $\theta$ is half the scattering angle, $\lambda$ is the wavelength of the incident X-rays, h and c° are Planck's constant and the speed of light and $E$ is the energy of the incident X-ray photon. Equation (2.3) is convenient for variable-wavelength measurements such as energy-dispersive X-ray diffraction (frequently referred to as the EDXD method; see, for example, [4]). Since the phase factor of scattered X-rays at the position $r$ is given by $\exp(-\mathrm{i}Q \cdot r)$, the amplitude of the scattered X-rays, $A(Q)$, is expressed in the static approximation as follows:

$$A(Q) = \sum_k f_k(Q) \exp(-\mathrm{i}Q \cdot r_k), \tag{2.4}$$

where $f_k(Q)$ is the atomic scattering factor for atom $k$ located at position $r_k$. Thus, the coherent X-ray scattering intensity, $I^{\mathrm{coh}}(Q)$, can be written as follows:

$$I^{\mathrm{coh}}(Q) = \langle A(Q)A^*(Q)\rangle = \Big\langle \sum_j \sum_k f_k(Q) \exp\Big[-\mathrm{i}Q\cdot(r_j - r_k)\Big]\Big\rangle \tag{2.5}$$

Here $\langle\cdot\rangle$ denotes the statistical average.

In non-crystalline systems without long-range atomic periodicity, the summation in (2.5) may be well approximated by the average value of the positional correlation over all orientations. Due to the spherical symmetry in a rather homogeneous system, the functions $f(Q)$ and $I^{\mathrm{coh}}(Q)$ depend only upon the magnitude of $Q$ (see, for example, [2]). It is noted, however, that the monotonic decrease in $f(Q)$ with increasing $Q$ is attributed to the intra-atomic interference effect, which is almost independent of the atomic distribution within the scattering process. Therefore, $I^{\mathrm{coh}}(Q)$ is quite likely normalized by removing this $Q$-dependence from the $f_k(Q)$ term. Then, the structure factor, which is directly related to the RDF in non-crystalline system, can be defined. For simplicity, consider at first a non-crystalline system containing only one kind of atom. Equation (2.5) reduces to the following form:

$$I^{\mathrm{coh}}(Q) = f^2(Q)\Big\langle \sum_j \sum_k \exp\Big[-\mathrm{i}Q\cdot(r_j - r_k)\Big]\Big\rangle. \tag{2.6}$$

Excluding the forward-scattering term, the structure factor, $S(Q)$, often referred to as the interference function, can be written as follows:

$$S(Q) = \frac{1}{N}\Big\langle \sum_j \sum_k \exp\Big[-\mathrm{i}Q\cdot(r_j - r_k)\Big]\Big\rangle - N\delta_{Q,0}, \tag{2.7}$$

where $N$ is the total number of atoms in a system with volume $V$ and the $\delta_{Q,0}$ term corresponds to the intensity at $Q = 0$. The contribution from the $\delta_{Q,0}$ term is frequently neglected in practical calculations, because its physical significance is limited to an extremely narrow region near $Q = 0$. $\rho(r)$ is expressed by the following:

$$\rho(r) = \rho_\circ g(r) = \frac{1}{N}\Big\langle \sum_j \sum_k \delta\Big[r - (r_j - r_k)\Big]\Big\rangle - N\delta(r). \tag{2.8}$$

By using the relation $\rho(r) = \rho_\circ[g(r) - 1] + \rho_\circ$, (2.8) can be re-written as follows:

$$\frac{1}{N}\Big\langle \sum_j \sum_k \delta\Big[r - (r_j - r_k)\Big]\Big\rangle - \rho_\circ = \rho_\circ[g(r) - 1] + \delta(r). \tag{2.9}$$

By applying Fourier transformation to (2.9), the following equation is obtained:

$$\frac{1}{N}\Big\langle \sum \sum \exp\Big[-\mathrm{i}Q\cdot(r_j - r_k)\Big]\Big\rangle - N\delta_{Q,0} = 1 + \rho_\circ \int [g(r) - 1]\exp(-\mathrm{i}Q\cdot r)\mathrm{d}r. \tag{2.10}$$

Here, the following relations are also used:

$$\frac{1}{V}\int \exp(-\mathrm{i}Q\cdot r)\mathrm{d}r = 1+\delta_{Q,0}, \tag{2.11}$$

$$\int \delta(r)\exp(-\mathrm{i}Q\cdot r)\mathrm{d}r = 1, \tag{2.12}$$

$$\int \delta(r-r')\exp(-\mathrm{i}Q\cdot r')\mathrm{d}r' = \exp(-\mathrm{i}Q\cdot r). \tag{2.13}$$

Thus, we can now obtain the following fundamental relation between the structure factor (or interference function) obtained directly from a diffraction experiment and the RDF for a non-crystalline system containing a single component:

$$S(Q) = 1+\rho_\circ\int\Big[g(r)-1\Big]\exp(-\mathrm{i}Q\cdot r)\mathrm{d}r. \tag{2.14}$$

The following well-known equation for estimating the reduced RDF data of $G(r)$ is easily derived from (2.14):

$$G(r) = 4\pi r\Big[\rho(r)-\rho_\circ\Big] = \frac{2}{\pi}\int_0^\infty Q\Big[S(Q)-1\Big]\sin(Q\cdot r)\mathrm{d}Q. \tag{2.15}$$

## 2.1 Partial Structure Function Analysis

The RDF analysis and its interpretation are more complicated in a system containing more than two kinds of components. However, when we introduce the compositionally averaged functions expressed by the following equations, a similar approach to that for the simple one-component system may be applicable:

$$\overline{\rho}(r) = \sum_{i=1}^{n}\sum_{j=1}^{n} c_i f_i f_j \rho_{ij}(r)/\langle f\rangle^2, \tag{2.16}$$

$$\langle f\rangle^2 = (\sum_{i=1}^{n} c_i f_i)^2, \tag{2.17}$$

$$\langle f^2\rangle = \sum_{i=1}^{n} c_i f_i^2, \tag{2.18}$$

where $c_i$ is the atomic fraction of $i$-type atom and $\rho_{ij}(r)$, is generally called the partial radial density function, and corresponds to the number of $i$-type atoms found at a radial distance of $r$ from a $j$-type atom at the origin. Here $Q$-dependence of the atomic scattering factor, $f(Q)$, is excluded again for simplification. Equation (2.16) implies that the average radial density function $\overline{\rho}(r)$, of a multi-component, non-crystalline system could be given by the summation of the partial radial density function $\rho_{ij}(r)$, with a weighting factor using the atomic scattering factor and concentration. By using the relations given by (2.16)–(2.18), the coherent X-ray scattering intensity per atom, $I_{\mathrm{a}}^{\mathrm{coh}}(Q)$, and the structure factor, $a(Q)$, often called the total structure factor, for a non-crystalline system containing more than two kinds of atoms are expressed as follows:

$$I_{\mathrm{a}}^{\mathrm{coh}}(Q) = \langle f^2 \rangle + \langle f \rangle^2 \int_0^\infty 4\pi r^2 \Big[\rho(r) - \rho_\circ\Big] \frac{\sin(Q \cdot r)}{Q \cdot r} \mathrm{d}r, \tag{2.19}$$

$$a(Q) = \Big[I_{\mathrm{a}}^{\mathrm{coh}}(Q) - \Big(\langle f^2 \rangle - \langle f \rangle^2\Big)\Big] \Big/ \langle f \rangle^2. \tag{2.20}$$

The following common relation, similar to (2.15), may also be obtained:

$$\overline{G}(r) = 4\pi r \Big[\overline{\rho}(r) - \rho_\circ\Big] = \frac{2}{\pi} \int_0^\infty Q \Big[a(Q) - 1\Big] \sin(Q \cdot r) \mathrm{d}Q. \tag{2.21}$$

Equation (2.21) provides the fundamental tool for extracting the information about the atomic-scale structure from measured X-ray scattering intensity data of non-crystalline systems including more than two components, although the information about $\overline{G}(r)$ cannot be used to describe completely the positions and chemical identities of the constituents. For this purpose, knowledge of the structure of individual pairs, such as $\rho_{ij}(r)$, is undoubtedly required. Such information is often called partial structure functions, and they are probably the unique items for understanding various characteristic properties of multi-component, non-crystalline materials at a microscopic level.

Since the surroundings of each atom (as also represented by the environmental structure around a specific component) in non-crystalline systems generally differ from those of other atoms, as easily seen in Fig. 2.2, the interpretation of RDF for the multi-component case is more complicated. In this respect, the partial RDFs for the individual pairs of chemical constituents are of particular importance and almost the only items for quantitatively describing the atomic-scale structure of multi-component, non-crystalline systems. The partial radial density function, $\rho_{ij}(r)$, can mathematically defined by

$$\rho_{ij}(r) = c_j \rho_\circ g_{ij}(r) = \frac{1}{N_i} \sum_{\alpha}^{N_i} \sum_{\beta}^{N_j} \delta\Big[r - (r_\alpha - r_\beta)\Big] - \delta_{\alpha\beta}\delta(r), \tag{2.22}$$

where $N_i$ is the number of $i$-type atoms in the volume $V$ and $g_{ij}(r)$ is the so-called partial pair distribution function. The summation of $i$ and $j$ is over the chemical constituents, whereas the summation of $\alpha$ and $\beta$ is over all atoms in the system of interest. The following relations are also usually employed:

$$c_i \rho_{ij}(r) = c_j \rho_{ji}(r), \tag{2.23}$$

$$N_i \rho_{ij}(r) = N_j \rho_{ji}(r). \tag{2.24}$$

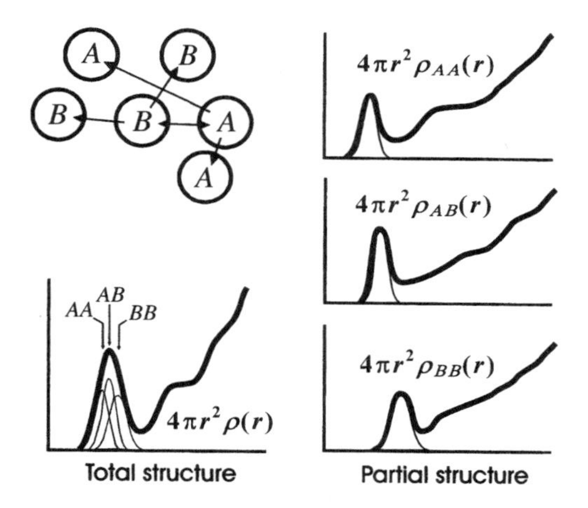


**Fig. 2.2.** Schematic for the partial distribution functions in a binary system

Therefore it follows immediately that $g_{ij}(r) = g_{ji}(r)$. Since the long-range atomic correlation disappears in non-crystalline systems, the probability of finding a pair of atoms, of course, approaches the average value with increasing distance. Thus,

$$\rho_{ij}(r) \to N_j/V = c_j\rho_\circ, \qquad g_{ij}(r) \to 1 \qquad (r \to \infty). \tag{2.25}$$

Hence, the basic features of the partial RDFs themselves are very similar to those of the RDF for the simple case containing only one component. However, the number of partial RDFs, corresponding to the possible atomic pairs, drastically increases as the number of constituents in the system increases. It follows that there are $n(n+1)/2$ possible pairs in the system containing $n$ components. Thus, three partial RDFs in a binary system and six partial RDFs in a ternary system are required to describe completely the atomic-scale structure.

The partial RDF is known to connect with the partial structure factor (often called the partial interference function) of the corresponding pair correlation determined from the diffraction experiments using X-rays and neutrons. However, the definition of the partial structure factor is not unique, and we can find three different equations proposed by Faber and Ziman [5], Ashcroft and Langreth [6] and Bhatia and Thornton [7]. Although all three equations are connected to each other by simple linear relations, they are each characterized by relative merits and demerits (see, for example, [8]). For convenience, the essential points of the partial structure factors are given below, using the Faber–Ziman (hereafter referred to as FZ) form in the binary case as an example.

Let us consider a binary, non-crystalline system containing two types of atoms, $A$ and $B$. Thus three different partial RDFs, related to $\rho_{AA}(r)$, $\rho_{BB}(r)$

and $\rho_{AB}(r)$, are required to describe the structure of this system. The total number of atoms, $N$, in a volume, $V$, consists of $N_A$ and $N_B$, where $N_A$ and $N_B$ are the number of atoms $A$ and $B$. Then, the atomic fractions are defined by $c_A = N_A/N$ and $c_B = N_B/N$, respectively. The coherent X-ray scattering intensity, $I^{\rm coh}(Q)$, analogous to (2.5), may be written in the following form:

$$I^{\rm coh}(Q) = f_A^2 \left\langle \Big| \sum_{j=1}^{N_A} \exp(-{\rm i}Q \cdot r_{Aj}) \Big|^2 \right\rangle + f_B^2 \left\langle \Big| \sum_{k=1}^{N_B} \exp(-{\rm i}Q \cdot r_{Bk}) \Big|^2 \right\rangle$$
$$+\ 2 f_A f_B \Big\langle \sum_{j=1}^{N_A} \sum_{k=1}^{N_B} \exp\Big[-{\rm i}Q \cdot (r_{Aj} - r_{Bk})\Big] \Big\rangle. \tag{2.26}$$

The three double sums in (2.26) correspond to the partial structure factors of the respective atomic pairs, $A$–$A$, $B$–$B$ and $A$–$B$.

Faber and Ziman [5] used the following definition, analogous to (2.14), for the partial structure factors, $a_{ij}(Q)$:

$$a_{ij}(Q) = 1 + \rho_\circ \int \Big[g_{ij}(r) - 1\Big] \exp(-{\rm i}Q \cdot r){\rm d}r. \tag{2.27}$$

Excluding the forward-scattering term, the FZ partial structure factors can be given as follows:

$$a_{ij}(Q) = (c_i c_j)^{-1/2} \Big\{ (N_i N_j)^{-1/2} \Big\langle \sum_\alpha \sum_\beta \Big[ -{\rm i}Q \cdot (r_{i\alpha} - r_{j\beta}) \Big] \Big\rangle$$
$$- (N_i N_j)^{1/2} \delta_{Q,0} - c_j^{-1} \delta_{ij} + 1 \Big\}. \tag{2.28}$$

By using the FZ formulation, the coherent X-ray scattering intensity per atom is given in the following form:

$$I_{\rm a}^{\rm coh}(Q) = \Big(\langle f^2 \rangle - \langle f \rangle^2\Big) + \sum_i \sum_j c_i c_j f_i f_j a_{ij}(Q), \tag{2.29}$$

where

$$\langle f^2 \rangle = c_A f_A^2 + c_B f_B^2, \tag{2.30}$$

$$\langle f \rangle = c_A f_A + c_B f_B, \tag{2.31}$$

$$(\langle f^2 \rangle - \langle f \rangle^2) = c_A c_B (f_A - f_B)^2. \tag{2.32}$$

The quantity given by (2.32) is frequently called the Laue monotonic scattering term, attributed to the intensity arising only from the difference in the atomic scattering factors of the constituent atoms. The FZ total structure factor is then expressed by

$$a_{\mathrm{FZ}}(Q) = \left[ I_{\mathrm{a}}^{\mathrm{coh}}(Q) - \left( \langle f^2 \rangle - \langle f \rangle^2 \right) \right] \Big/ \langle f \rangle^2 \tag{2.33}$$

$$= \sum_i \sum_j c_i c_j \frac{f_i f_j}{\langle f \rangle^2} a_{ij}(Q). \tag{2.34}$$

These FZ expressions are used in (2.19) and (2.20).

In contrast to the proposal of Faber and Ziman [5], Ashcroft and Langreth (hereafter referred to as AL) [6] used a different expression for the partial structure factors, $S_{ij}(Q)$. Their definitions corresponding to the FZ case are as follows:

$$I_{\mathrm{a}}^{\mathrm{coh}}(Q) = \sum_i \sum_j (c_i c_j)^{1/2} f_i f_j S_{ij}(Q), \tag{2.35}$$

$$S_{\mathrm{AL}}(Q) = I_{\mathrm{a}}^{\mathrm{coh}}(Q) / \langle f^2 \rangle = \sum_i \sum_j (c_i c_j)^{1/2} \frac{f_i f_j}{\langle f^2 \rangle} S_{ij}(Q). \tag{2.36}$$

It may also be noted that these different sets of partial structure factors can be mutually transformed by means of the following linear equations:

$$S_{AA}(Q) = 1 + c_A \left[ a_{AA}(Q) - 1) \right], \tag{2.37}$$

$$S_{BB}(Q) = 1 + c_B \left[ a_{BB}(Q) - 1) \right], \tag{2.38}$$

$$S_{AB}(Q) = (c_A c_B)^{1/2} \left[ a_{AB}(Q) - 1) \right]. \tag{2.39}$$

With respect to the physical significance of the FZ and AL partial structure factors, a brief comment is given for convenience of further discussion. As easily seen in (2.33) and (2.36), different normalization has to be performed. The three partial structure factors in the FZ form vary around unity, while the AL partial structure factor of unlike atom pairs, $S_{AB}(Q)$, oscillates around zero in contrast to the variation around unity in the two partial structure factors of like-atom pairs, $S_{AA}(Q)$ and $S_{BB}(Q)$. The FZ form expresses mutually comparable quantities; for example, a substitutional alloy system in which a solute atom can replace a solvent atom without any constraint such as the change in volume. In that sense, the concentration dependence of the FZ partial structure factors corresponds to the deviation from ideal behavior where the concentration independence is recognized (see, for example, [9]). This is the reason for the fact that the FZ partial structure factors are relatively insensitive to the concentration in comparison with the AL partial structure factors. However, it should be kept in mind that the partial structure factors of both definitions should, in principle, be functions of the concentration.

## 2.2 Environmental Structure Function Analysis

The concept of the environmental structure function around a specific element is also quite useful for discussing the structure–property relationships of various materials of interest. This is particularly true in multi-component systems containing more than three elements, because the actual implementation of the respective partial functions from measured structure data is not a trivial task even for a binary system. The basic concept of the partial structure is perfectly unchanged in this environmental structure analysis. The idea of this data processing is to differentiate one element, for example, $A$, in a ternary system containing $A$, $B$ and $C$ elements and to estimate its local chemical environmental structure as a function of radial distance corresponding to the average atomic arrangements produced only from three partial functions of the $A$–$A$, $A$–$B$ and $A$–$C$ pairs, without complete separation into six partial functions of individual constituents [10]. The idea of the environmental structure analysis itself is not new, but successful results for complicated systems have been obtained only recently. For example, the usefulness of this data processing was well confirmed by the results obtained using the isotope substitution method of neutrons for solutions [11] as well as the AXS method for various amorphous alloys [12] and solutions [13]. It should also be noted that the environmental structure analysis around a specific element is applicable not only to non-crystalline systems but also to crystalline systems using the simple Fourier transformation, as clearly shown with some selected examples of ultrafine particles [14] and a high-temperature superconducting oxide [15].

Furthermore, environmental structure analysis has recently drawn much attention, because this data processing provides about an order of magnitude higher stability for the solutions than the direct AXS method [10, 16, 17]. In this case, we are not requested to separate all partial functions for individual pairs of the constituents, and it is very convenient for analyzing the structure of a non-crystalline system containing more than two elements. For this reason, the essential equations are given below using the AXS case as an example. When the incident energy is set to the close vicinity below the absorption edge, $E_{\mathrm{abs}}$, of the $A$-element, the anomalous dispersion phenomena become significant and thereby the variation between the reduced interference functions obtained from the measurements at two energies, $E_1$ and $E_2$ ($E_1 < E_2 < E_{\mathrm{abs}}$), is attributed only to the change in the real part of the anomalous dispersion term of the $A$-element. The following relation is readily obtained:

$$\Delta i_A(Q, E_1, E_2) = \sum_k \frac{c_k \Re[f_k(Q, E_1) + f_k(Q, E_2)]}{W(Q, E_1, E_2)}[a_{Ak}(Q) - 1], \tag{2.40}$$

$$W(Q, E_1, E_2) = \sum_k c_k \Re[f_k(Q, E_1) + f_k(Q, E_2)], \tag{2.41}$$

where $\Re$ denotes the real part of the values in the brackets. The terms in front of $[a_{Ak}(Q) - 1]$ are the so-called effective weighting factors. This corresponds to the two independent scattering measurements when changing the scattering factor of the $A$-element without any change in the structure and concentration. Then, the intensity variation can be estimated by simply subtracting the difference of the mean square of the atomic scattering factors at these two energies. It is also worth mentioning that variation detected in these two measurements yields the environmental structure function that represents the contribution from the structure-sensitive part related only to the $A$-element, for example, the average structure of $A$–$A$ and $A$–$B$ in a binary case and that of $A$–$A$, $A$–$B$, and $A$–$C$ in a ternary one. The environmental interference function is related to the pair distribution function as follows:

$$\Delta i_A(Q, E_1, E_2) = \sum_k \frac{c_k \Re[f_k(Q, E_1) + f_k(Q, E_2)]}{W(Q, E_1, E_2)} \left(\frac{1}{Q}\right) \times \int_0^\infty 4\pi r \rho_\circ [g_{Ak}(r) - 1] \sin(Q \cdot r) \mathrm{d}r. \tag{2.42}$$

The reduced environmental RDF, $G_A(r)$, that presents the atomic environment around a specific element, $A$, is obtained in the following form:

$$G_A(r) = \sum_k \frac{c_k \Re[f_k(Q, E_1) + f_k(Q, E_2)]}{W(Q, E_1, E_2)} \{4\pi r \rho_\circ [g_{Ak}(r) - 1]\} = \frac{2}{\pi} \int_0^\infty Q \Delta i_A(Q) \sin(Q \cdot r) \mathrm{d}Q. \tag{2.43}$$

The effective weighting factors for the RDFs are usually approximated to an average value in the range of $Q$, and then the environmental RDF can also be written as follows:

$$4\pi r^2 \rho_A(r) = 4\pi r^2 \rho_\circ + \frac{2r}{\pi} \int_0^\infty Q \Delta i_A(Q) \sin(Q \cdot r) \mathrm{d}Q. \tag{2.44}$$

## References


1. T.L. Hill: *Statistical Mechanics* (McGraw-Hill, New York 1956)
2. B.E. Warren: *X-ray Diffraction* (Addison-Wesley, Reading 1969)
3. C.N.J. Wagner: *Liquid Metals, Chemistry and Physics*, ed. by S.Z. Beer (Marcel-Dekker, New York 1972) pp. 258
4. T. Egami in: *Glassy Metals*, ed. by H.J. Güntherodt and H. Beck (Springer, Berlin, Heidelberg, New York 1981) pp. 25
5. T.E. Faber and J.M. Ziman: Philos. Mag., **11**, 153 (1965)
6. N.W. Ashcroft and D.C. Langreth: Phys. Rev., **159**, 500 (1967)
7. A.B. Bhatia and D.E. Thornton: Phys. Rev., B **2**, 3004 (1970)

8. Y. Waseda: *The Structure of Non-Crystalline Materials* (McGraw-Hill, New York 1980)
9. N.C. Halder and C.N.J. Wagner: J. Chem. Phys., **47**, 4385 (1967)
10. Y. Waseda: *Novel Application of Anomalous X-ray Scattering for Structural Characterization of Disordered Materials* (Springer, Berlin, Heidelberg, New York 1984)
11. G.W. Neilson and J.E. Enderby: Proc. R. Soc. Lond., A **390**, 353 (1983)
12. E. Matsubara and Y. Waseda: Mater. Trans. JIM, **36**, 883 (1995)
13. K.F. Ludwig Jr., W.K. Warburton and A. Fontaine: J. Chem. Phys., **87**, 620 (1987)
14. E. Matsubara, K. Okuda, Y. Waseda and T. Saito: Z. Naturforsch., **47a**, 1023 (1992)
15. K. Sugiyama and Y. Waseda: J.Mater. Sci., **7**, 450 (1988)
16. P.H. Fuoss, P. Eisengerger, W.K. Warburton and A. Bienenstock: Phys. Rev. Lett., **46**, 1537 (1981)
17. R.G. Munro: Phys. Rev., B **25**, 5037 (1982)

# 3. Nature of Anomalous X-ray Scattering and Its Application to the Structural Analysis of Crystalline and Non-crystalline Systems

Each atom has its own absorption edges for X-rays at certain characteristic energies. Such edges represent the threshold excitation energies above which an inner electron can be ejected into the continuum states (see, for example, [1]). These X-ray absorption phenomena include predominantly the excitation of K-shell or L-shell electrons, and thus our attention focuses mainly upon the K-absorption edge or the L-absorption edge. In conventional X-ray diffraction analysis, we generally choose the energy (or wavelength, hereafter the term of energy is used) of incident X-rays away from such an absorption edge of the constituent elements, and the energy independence is then well accepted for the so-called atomic scattering factor, $f(Q)$, given by simple potential scattering theory.

On the other hand, when the energy of the incident X-ray beam is close to such an absorption edge of the constituent elements, $f(Q)$ becomes complex and can be expressed in the following form:

$$f(Q,E) = f^{\circ}(Q) + f'(E) + \mathrm{i}f''(E), \tag{3.1}$$

where $Q$ and $E$ are the wave vector and the incident X-ray energy, respectively. The first term of (3.1) corresponds to the normal atomic scattering factor given by the Fourier transform of the electron density in an atom, for radiation at an energy much higher than any absorption edge, and $f'$ and $f''$ are the real and imaginary components of the so-called anomalous dispersion term.

Since the spatial distribution of inner electrons is considerably smaller than the magnitude of the X-ray wavelength, the dipole approximation $[\exp(-\mathrm{i}Q \cdot r) \fallingdotseq 1]$ is well accepted, and thus the $Q$-dependence of the anomalous dispersion factors $f'$ and $f''$ can be ignored. However, such $Q$-dependence may be required for a discussion of $f'$ and $f''$ near the absorption edges of M and N series (see, for example, [2]).

The anomalous dispersion terms of $f'$ and $f''$ arising from anomalous (resonance) scattering depend upon the incident X-ray energy, and their variation as a function of energy is illustrated in Fig. 3.1 using the K-absorption edge of an Fe atom as an example. The salient features are as follows: The imaginary part of $f''$ is positive and distinguished only on the higher-energy side of the absorption edge. On the other hand, the real part of $f'$ indicates a sharp negative peak at the absorption edge, and its width is typically 50 eV

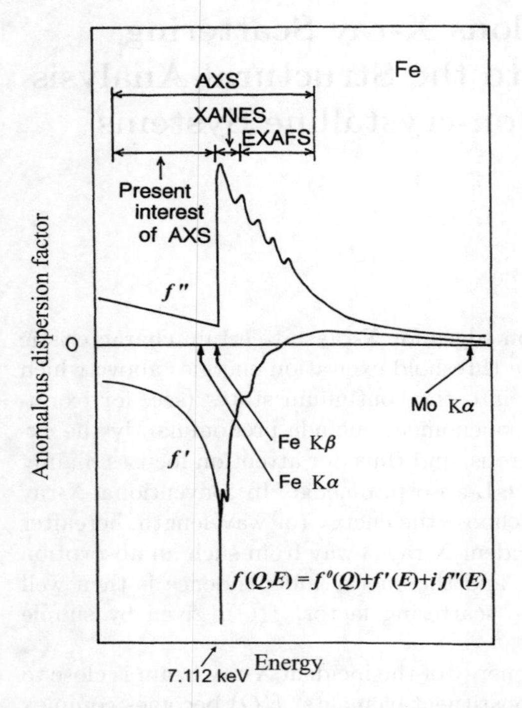


**Fig. 3.1.** Energy variation of anomalous dispersion factors for Fe

at half maximum. As easily seen in Fig. 3.1, the component $f'$ exists on either side of the edge, and the maximum effect at the energy rapidly reduces and approaches an approximately constant level for energies a few hundred eV away from the edge.

The dispersion behavior of the imaginary part, $f''$, follows simple absorption phenomena. That is, the absorption edge corresponds to the threshold energy (frequency) above which an inner electron can be ejected into the continuum states. This process can take place only for the case in which the incident X-ray energy is equal to or greater than that of the absorption edge. The real part of $f'$ is generally observed as a phase difference in the X-ray optics, and its dispersion behavior is not independent of that of $f''$. Such a relation is represented by the so-called Kramers–Krönig transform [1]. It is frequently called the "dispersion relation" and can be written in the following generalized form:

$$f'(E) = \frac{2}{\pi}\int_0^\infty \frac{f''(E)E'}{E^2 - E'^2}\,\mathrm{d}E', \tag{3.2}$$

where $E$ denotes the photon energy. This relation suggests that knowledge of $f''(E)$ over a sufficiently wide energy region provides a method for evaluating the dispersion behavior of the real part, $f'(E)$.

As shown in Fig. 3.1, the energies of some characteristic X-rays, such as Fe-K$\alpha$(6.404 keV) and Fe-K$\beta$(7.057 keV), are located near an absorption edge (7.112 keV) of the Fe atom, and the anomalous dispersion factors become quite sizable. For example, in the Fe atom, $f' = -2.10$ and $f'' = 0.57$ for the Fe-K$\alpha$ radiation, and the change of $f'$ with respect to $f^\circ(Q)$ corresponds to about 8%. Thus, the measurements of X-ray scattering intensity at an energy near the absorption edge of the constituent element gives an additional piece of information about the structure of the desired system.

In order to facilitate the understanding of the anomalous dispersion effect for differentiating between two components, let us consider a sample containing Fe and Ni. Since the normal X-ray scattering factor is known to be proportional to the atomic number, the scattering ability may be approximated to be 26 for Fe and 28 for Ni. The numerical values of the real part of $f'$ for two elements, Fe and Ni, are summarized in the second and third column of Table 3.1 for three characteristic X-rays (Mo-K$\alpha$, Fe-K$\alpha$ and Fe-K$\beta$) together with those for the case in which the incident energy is tuned at 7.110 keV. The measurement using Mo-K$\alpha$, corresponding to the case at an energy away from the edge, shows no appreciable difference in intensity. The factor of 2 is unchanged. However, the anomalous dispersion effect becomes significant when using Fe-K$\alpha$ and Fe-K$\beta$, suggesting that the difference detected in intensity increases by a factor of 3 and 5, respectively (see Table 3.1). The factor drastically increases up to 8 when the incident energy is tuned to the value of 7.110 keV, that is, very close to the absorption edge of Fe. As clearly demonstrated in Table 3.1, sufficient atomic sensitivity can be obtained even for substances containing near-neighbor elements in the periodic table, simply by changing the incident energy of the X-rays.

**Table 3.1.** Real part of the anomalous dispersion factors of Fe and Ni and scattering ability at energies of Mo-K$\alpha$, Fe-K$\alpha$, Fe-K$\beta$ and 7.110 keV. $f^\circ_{\mathrm{Fe}} = 26$ and $f^\circ_{\mathrm{Ni}} = 28$ at $\sin\theta/\lambda = 0$

| Energy (keV) | $f'_{\mathrm{Fe}}$ | $f'_{\mathrm{Ni}}$ | $f^\circ + f'$ | Difference |
|---|---|---|---|---|
| 17.480 (Mo-K$\alpha$) | 0.3 | 0.3 | Fe $26 + 0.3 = 26.3$ | $\fallingdotseq$ 2 |
| | | | Ni $28 + 0.3 = 28.3$ | |
| 6.404 (Fe-K$\alpha$) | −2.1 | −1.3 | Fe $26 - 2.1 = 23.9$ | $\fallingdotseq$ 3 |
| | | | Ni $28 - 1.3 = 26.7$ | |
| 7.057 (Fe-K$\beta$) | −5.2 | −2.3 | Fe $26 - 5.2 = 20.8$ | $\fallingdotseq$ 5 |
| | | | Ni $28 - 2.3 = 25.7$ | |
| 7.110 | −7.9 | −1.7 | Fe $26 - 7.9 = 18.1$ | $\fallingdotseq$ 8 |
| | | | Ni $28 - 1.7 = 26.3$ | |

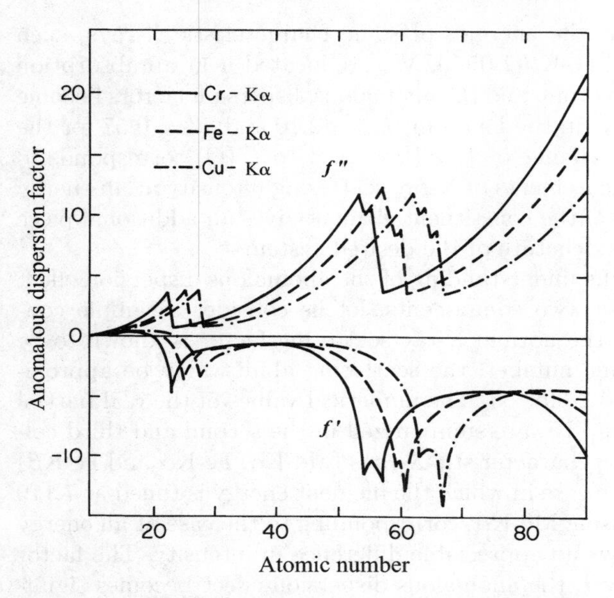


**Fig. 3.2.** Anomalous dispersion factors of various elements for the characteristic Kα radiation of Cr, Fe and Cu as a function of atomic number [4]

As shown in Fig. 3.2, the anomalous dispersion terms for some characteristic X-rays indicate a discontinuous variation when plotted against the atomic number [4]. It is also worth noting that the change in the anomalous dispersion terms, $f'$ and $f''$, is quite distinct for 3d transition metals and 4f rare-earth metals. This implies that the anomalous X-ray scattering (AXS) method could be applicable to samples containing 3d transition metals or 4f rare-earth metals using the characteristic radiation. It is also worth mentioning that for most of the elements with rregard to AXS the real part of the anomalous dispersion term, $f'$, is typically 15–20% of the standard atomic scattering factor, $f^{\circ}(Q)$, at the K-shell absorption edge, and $f'$ appears to be substantially larger value (over 30%) at the L-shell absorption edge, as exemplified by the case for the Cs atom in Fig. 3.3 (see, for example, [3]). However, the energy of the characteristic radiation is often not close enough to the absorption edges of the constituent elements. For this reason, efficient use of the AXS effect can be attained for only a relatively small number of systems, as long as we use the characteristic radiation produced from commercial X-ray targets. With respect to this subject, the availability of intense white (continuous) X-rays from a synchrotron radiation source has greatly improved both the acquisition and quality of the AXS data by enabling the use of energy where the anomalous dispersion effect is the greatest. Figure 3.4 shows a typical energy spectrum of a synchrotron radiation source available in the Photon Factory, Institute of Materials Structure Science, High Energy

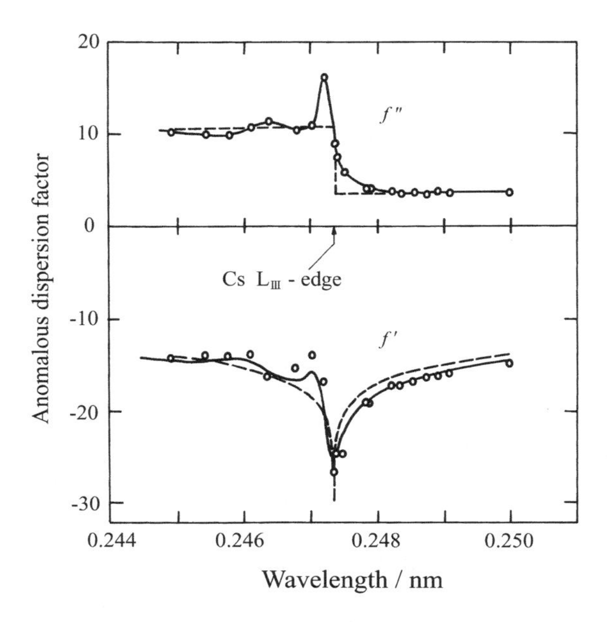


**Fig. 3.3.** Anomalous dispersion factors of Cs measured by the single-crystal diffraction measurement (*circles*) and calculation (*dashed lines*) near the $L_{III}$ absorption edge [3]

Accelerator Research Organization, Tsukuba, Japan [5]. At this facility, the energies of interest can be tuned, for example, by using a Si(111) double-crystal monochromator with an optimum energy resolution of about 5 eV at 10 keV. The effect of the higher harmonics of the Si(333) reflection is usually reduced by intentionally de-tuning the second Si crystal monochromator with a piezo-electric crystal device, although about one-fifth of the intensity of the first-order reflection is lost.

## 3.1 Application to Qualitative and Quantitative Powder Diffraction Analysis

X-ray diffraction is a very useful tool for the determination of the fine structure represented by the atomic arrangement of matter in a variety of states. In addition, other uses have been developed for diverse problems such as chemical analysis and particle-size determination. Here, some essential points for the quantitative analysis of a sample containing more than two phases and the determination of site occupancy of a particular element in poly-crystalline materials are presented with special reference to the use of the AXS effect.

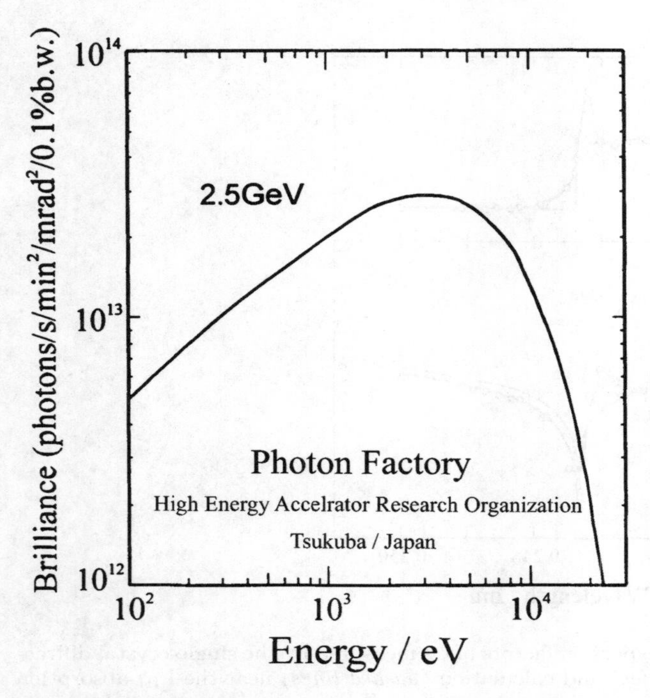


**Fig. 3.4.** Typical energy spectrum of a synchrotron radiation source (Tsukuba, Japan)

Quantitative analysis by diffraction is based on the fact that the diffraction intensity of a particular phase in a mixed sample depends upon the concentration of that phase in the mixture. There are number of papers in the literature on this subject and its application to the analysis of crystalline mixtures [6,7]. According to these studies, the diffracted intensity for the $i$-th reflection from the $j$-th phase at a wave length $\lambda$ (or an energy $E$) may be expressed as follows:

$$I_{ij\lambda} = \frac{K I_\lambda^\circ \rho_{ij} |F_{ij\lambda}|^2 \lambda^3 (Lp_\lambda) g_j}{\overline{\mu}_\lambda^* V_j^2 \rho_j}. \tag{3.3}$$

The symbols have their usual crystallographic meanings:

| | | |
|---|---|---|
| $I_\lambda^\circ$ | : | incident beam intensity; |
| $K$ | : | scaling factor depending upon the experimental conditions; |
| $\overline{\mu}_\lambda^*$ | : | averaged mass-absorption coefficient for the mixture; |
| $V_j$ | : | unit-cell volume of the $j$-th component; |
| $Lp_\lambda$ | : | the Lorentz-polarization factor depending upon the instrumental conditions; |

$g_j$ : the fractional weight of the $j$-th sample;
$\rho_j$ : the density of the $j$-th sample;
$\rho_{ij}$ : the multiplicity factor of the $i$-th reflection diffracted from the $j$-th component;
$F_{ij\lambda}$ : the scattering factor of the $i$-th reflection diffracted from the $j$-th component with the wavelength $\lambda$.

Equation (3.3) applies only to an ideally diffracting powder in which the individual crystalline phases are randomly oriented (no texture) and are sufficiently small that only the macro-absorption effects are considered. For simplification, let us consider a two-phase mixture ($j = A$ and $B$) at two energies ($\lambda = E_1 = \alpha$ and $E_2 = \beta$) and omit the subscript "$i$" with respect to (3.3).

When the intensity profiles of a sample are measured at two energies near the absorption edge of a specific element and this element is included only in the crystalline component of $A$, the anomalous dispersion effect arising from the component $B$ (matrix) appears to be insignificant, namely $|F_{B\alpha}| = |F_{B\beta}|$. In other words, the energy dependence of the intensities should be detectable only around the reflections originating from component $A$. For this reason, the ratio $I_\alpha^\circ / I_\beta^\circ$ can be estimated by comparing the two integrated intensity profiles of component $B$ (matrix) with the aid of the calculated Lorentz-polarization factors and mass-absorption coefficient. Using this ratio for a certain scaling factor, the value of $R_{j\lambda}$ defined by the following equation can readily be obtained:

$$R_{i\lambda} = \frac{K I_\alpha^\circ p_j |F_{j\lambda}|^2 \alpha^3 (Lp\alpha) g_j}{V_j^2 \rho_j}. \tag{3.4}$$

For a pure standard of component $A$, the following useful relation expresses the energy dependence of the intensity profile:

$$\Delta I_{\mathrm{element}}^\circ = K I_\alpha^\circ p_1 \alpha^3 (Lp\alpha)(|F_{1\alpha}|^2 - |F_{1\beta}|^2) \Big/ V_1^2 \rho_1. \tag{3.5}$$

The weight fraction of component $A$ in a mixture of two components, $A$ and $B$ in the present case, can be calculated as follows:

$$g_A = \Delta I_{\mathrm{element}} / \Delta I_{\mathrm{element}}^\circ. \tag{3.6}$$

Thus, quantitative determination of the component $A$ results from the different patterns measured at the two energies $E_1 = \alpha$ and $E_2 = \beta$.

In a mixture containing more than two phases, we frequently find difficulty in identifying the individual chemical constituents by the conventional X-ray powder diffraction method. For example, this is certainly the case for a mixture of copper sulfide ($Cu_2S$) and ferrite ($Fe_2O_3$) components discharged from a copper smelting process. Because there are insufficient differences in the X-ray diffraction intensities of two elements (Cu and Fe in the present case), of nearly the same atomic number in the periodic table, they cannot be detected. In order to reduce such difficulty, for example, Nichols et al. [6] proposed a procedure using the difference in the diffraction intensity profiles

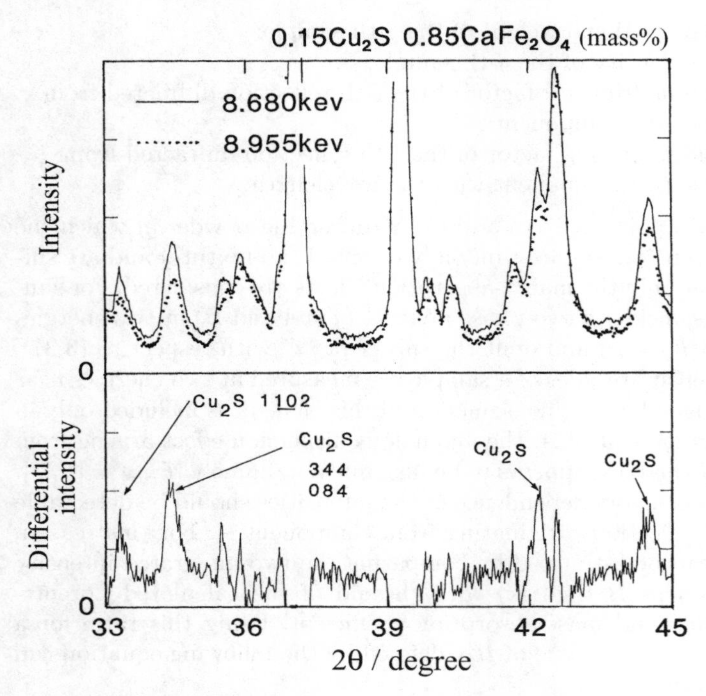


**Fig. 3.5.** Differential intensities (*bottom*) of $Cu_2S$ obtained from the intensity data (*top*) measured at the two energies 8.680 keV (*solid line*) and 8.955 keV (*dotted line*), corresponding to energies of 300 and 25 eV below the Cu-K absorption edge [8]

arising from the anomalous dispersion effect of a particular element of one phase in the mixture. Figure 3.5 shows the energy variation of pure $Cu_2S$ obtained from the measurements at energies of 8.680 and 8.955 keV, which correspond to energies of 300 and 25 eV below the Cu-K absorption edge (8.980 keV) [8]. Here the difference, $\Delta I_{\mathrm{Cu}}$, is attributed to the anomalous dispersion effect of copper. Using this profile as the standard, a quantitative analysis of $Cu_2S$ can be readily carried out by means of the present AXS method.

There is a long-standing tradition of cation site preferences in ferrite spinels and metal ordering in the alloy systems consisting of similar atomic numbers. Several studies have clearly indicated that the AXS method can provide an elemental contrast, and this method is considered quite useful. The fundamentals for determining the cation distribution in crystalline materials are presented below. The diffracted intensity of the *hkl* reflection from a crystalline powder in a reflection geometry may be expressed in the following form in the way similar to (3.3) [9]:

$$I_{hkl}(E) = KpI_{\lambda}^{\circ}\lambda^3 PLA\frac{|F_{hkl}(E)|^2}{v_c^2}. \tag{3.7}$$

Symbols denote the following usual meaning:

| | | |
|---|---|---|
| $I_{hkl}$ | : | integrated intensity; |
| $I^{\circ}_{\lambda}$ | : | incident beam intensity; |
| $K$ | : | scaling factor depending upon the experimental conditions; |
| $\lambda$ | : | wavelength; |
| $p$ | : | multiplicity factor; |
| $P$ | : | polarization factor; |
| $L$ | : | Lorentz factor; |
| $A$ | : | absorption factor; |
| $v_c$ | : | volume of unit cell; |
| $F_{hkl}$ | : | crystallographic structure factor for the unit cell. |

$F_{hkl}$ is given by the following expression:

$$F_{hkl} = \Sigma f_j \exp[B_j(\sin\theta/\lambda)^2] \exp[2\pi \mathrm{i}(hx_j + ky_j + lz_j)], \tag{3.8}$$

where $(x_j, y_j, z_j)$ and $B_j$ are the fractional coordinates and the isotropic temperature factor, respectively, for the $j$-th component. The summation extends over all the atoms in the unit cell.

In the case of spinel structures, as exemplified by zinc ferrite ($\mathrm{Zn_xFe_{3-x}O_4}$) [10], the so-called crystallographic structure factor can be easily rewritten using the following three terms for tetrahedral A-site, octahedral B-site and the positions occupied by oxygen (O):

$$\begin{aligned} F_{hkl} = {} & f_{\mathrm{A}} \exp[B_{\mathrm{A}}(\sin\theta/\lambda)^2] \sum_{\mathrm{A}} \exp[2\pi \mathrm{i}(hx_j + ky_j + lz_j)] \\ & + f_{\mathrm{B}} \exp[B_{\mathrm{B}}(\sin\theta/\lambda)^2] \sum_{\mathrm{B}} \exp[2\pi \mathrm{i}(hx_j + ky_j + lz_j)] \\ & + f_{\mathrm{O}} \exp[B_{\mathrm{O}}(\sin\theta/\lambda)^2] \sum_{\mathrm{O}} \exp[2\pi \mathrm{i}(hx_j + ky_j + lz_j)]. \end{aligned} \tag{3.9}$$

For convenience, Fig. 3.6 shows a schematic for a spinel structure indicating two crystallographic non-equivalent cation sites.

For example, when we introduce the relation $x = \mathrm{Zn_B}/\mathrm{Zn_{total}}$, the scattering factors $f_{\mathrm{A}}$ and $f_{\mathrm{B}}$ are given as a function of $x$, as follows:

$$f_{\mathrm{A}} = (1 - x) f_{\mathrm{Zn}} + x f_{\mathrm{Fe}}, \tag{3.10}$$

$$f_{\mathrm{B}} = \frac{x}{2} f_{\mathrm{Zn}} + \frac{2 - x}{2} f_{\mathrm{Fe}}. \tag{3.11}$$

It should be noted that $x$ is frequently called an inversion parameter, and $x = 0.0$ and $x = 1.0$ correspond to the cation distribution of all zinc cations occupying the tetrahedral A-site and the octahedral B-site, respectively.

When the intensities are measured at two energies, $E_1$ and $E_2$, in the close vicinity of the absorption edge of a specific element such as Zn in the present case, the following simple but useful relation can readily be obtained:

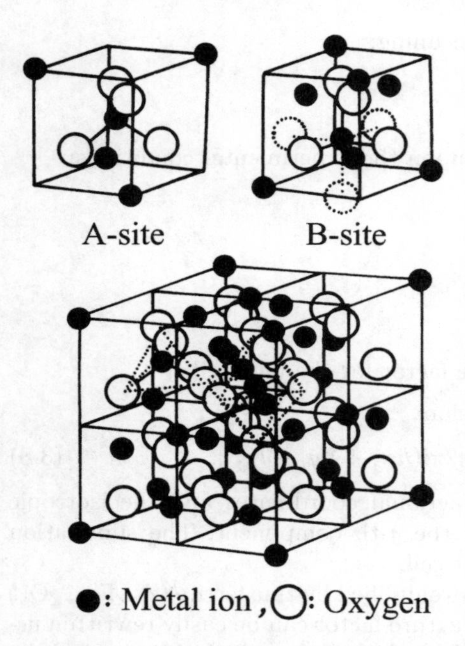


**Fig. 3.6.** Schematic for a spinel structure showing two crystallographic non-equivalent cation sites

$$r_{hkl,\mathrm{exp}} = \frac{I_{hkl}(E_1)/(\lambda_{E_1}^3 P_{E_1} L_{E_1}) I_0(E_2)}{I_{hkl}(E_2)/(\lambda_{E_2}^3 P_{E_2} L_{E_2}) I_0(E_1)} = \frac{|F_{hkl}(E_1)|^2}{|F_{hkl}(E_2)|^2}. \tag{3.12}$$

The detected intensity variation with energy may be attributed to the reflections originating only from a specific element, because the anomalous dispersion effects attributed to other components, for example, Fe in the present case, appear to be insignificant in this energy region. Thus, by comparing the measured $r_{hkl,\mathrm{exp}}$ at two energies close to the edge with the values of $r_{hkl,\mathrm{cal}}$ calculated from the model cation distribution in the usual manner, the site occupancy of the desired element, Zn in the present case, can be quantitatively determined. To obtain a significant result in data processing, the following $R$-factor and the iterative procedure shown in Fig. 3.7 are usually employed [11]:

$$R_{\mathrm{AXS}} = \sum_{hkl} \frac{[r_{hkl,\mathrm{exp}} - r_{hkl,\mathrm{cal}}]^2}{\sigma_{hkl,\mathrm{exp}}}, \tag{3.13}$$

where $\sigma_{hkl,\mathrm{exp}}$ is the fractional weight of $(1 - r_{hkl,\mathrm{exp}})^{-2}$. The cation distribution can be obtained so as to minimize the factor $R_{\mathrm{AXS}}$ of measurements at energies close to the absorption edges of the desired element.

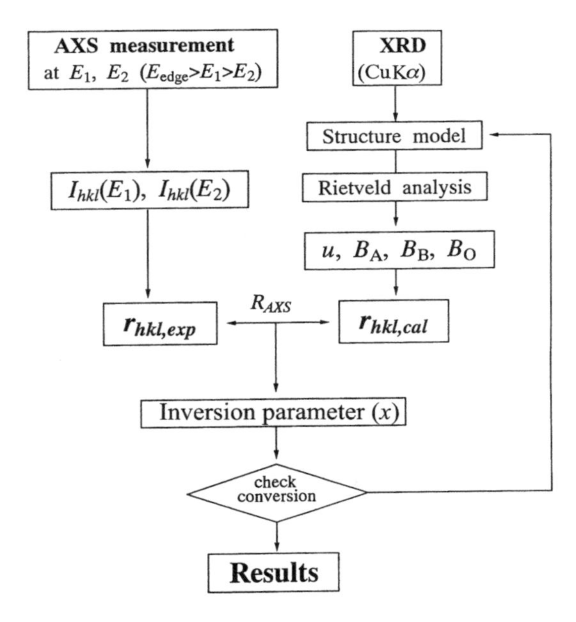


**Fig. 3.7.** Flow-chart of the iterative procedure for obtaining improved X-ray structural information from AXS measurements coupled with common Rietveld analysis [11]

## 3.2 Application to Radial Distribution Function Analysis for Non-crystalline Systems

When the full set of partial structure functions for the individual pairs of chemical constituents are obtained, we can have more than just one-dimensional information and thus better understand the structure–property relationships of non-crystalline systems in a more realistic way, including the compositional short-range order. With respect to this subject, efficient application of the AXS method is stressed when coupled with an intense white X-ray source such as synchrotron radiation. We give here some essential points for separating the so-called partial structure factors, which represent the Fourier transform of the partial radial distribution functions (RDFs), in a binary non-crystalline system using measured scattering intensity data.

First, it may be helpful to recall a few fundamental equations. The total structure factor, $a(Q)$, in terms of the Faber–Ziman (FZ) form, corresponding to the structurally sensitive part in (2.29), can be given by the summation of three partial structure factors of two like-atom pairs ($A$–$A$ and $B$–$B$) and one unlike-atom pair ($A$–$B$):

$$a(Q) = w_{AA}a_{AA}(Q) + w_{BB}a_{BB}(Q) + 2w_{AB}a_{AB}(Q), \tag{3.14}$$

$$w_{ij} = c_i c_j f_i f_j / \langle f \rangle^2, \tag{3.15}$$

where $c$ and $f$ denote the concentration and the atomic scattering factor, respectively. Regarding the partial RDFs, it is also customary to use the following relation:

$$G(r) = w_{AA}G_{AA}(r) + w_{BB}G_{BB}(r) + 2w_{AB}G_{AB}(r). \tag{3.16}$$

The separation of three individual structural functions ($a_{ij}(Q)$ or $G_{ij}(r)$) is the present objective. We will discuss here only the partial structure factors, $a_{ij}(Q)$, because $G_{ij}(r)$ is straightforward as given by the common Fourier transformation in the manner of (2.27).

The coefficient $w_{ij}$ in (3.15), frequently called the weighting factor, depends upon the concentrations and the atomic scattering factors. Thus the individual partial structure factors can be estimated only by making available at least three independent scattering experiments for which the weighting factors are varied without any change in the RDFs. For example, when the scattering ability is changed in the component $A$, the following matrix form can be readily obtained:

$$\begin{bmatrix} a_{AA}(Q) \\ a_{BB}(Q) \\ a_{AB}(Q) \end{bmatrix} = \begin{bmatrix} c_A^2 f_A^2 & c_B^2 f_B^2 & 2c_A c_B f_A f_B \\ c_A^2 (f_A^+)^2 & c_B^2 f_B^2 & 2c_A c_B f_A^+ f_B \\ c_A^2 (f_A^\#)^2 & c_B^2 f_B^2 & 2c_A c_B f_A^\# f_B \end{bmatrix}^{-1} \begin{bmatrix} a(Q) \\ a^+(Q) \\ a^\#(Q) \end{bmatrix}. \tag{3.17}$$

This has been done using several methods in the past. They can be classified as follows:

(i) Three different radiation techniques using X-rays, neutrons and electrons.
(ii) Isotope substitution technique for neutrons where the scattering ability of the component is varied using different isotopes.
(iii) Polarized neutron diffraction technique that is applicable only to magnetic materials.
(iv) Anomalous scattering technique for both X-rays and neutrons.

Of course, an assortment of the above techniques such as the combination of X-ray and neutron diffraction with the polarized neutron technique (see, for example, [12]) has been used in the literature. Table 3.2 is a summary of the various techniques for determining the partial structure factors in binary non-crystalline materials and the systems to which they were applied. The relative merit and demerit of each technique have already been discussed [4, 13] and will not be repeated here. Nevertheless, the following intrinsic nature is considered to be worthy of note.

Technique (i) requires samples different in size: bulk (mm) for neutron, foil (μm) for X-rays and thin film (nm) for electrons. In technique (ii), the structure is automatically assumed to remain identical upon substitution of

**Table 3.2.** Various methods for separating the partial structure functions in a binary non-crystalline system and their selected examples

| Combination of diffraction techniques | Examples of the results | | | | | |
|---|---|---|---|---|---|---|
| (A) X-ray, neutron and electron | Glass | Pd–Si | [14], | Mn–Si | [15] |
| (B) Neutron (isotope substitution technique) | Glass | Cu–Zr | [16], | Fe–Ge | [17] |
| | Liquid | Cu–Sn | [18], | RbCl | [19] |
| (C) Neutron (polarized neutron technique) | Glass | Co–P | [20] | | |
| (D) X-ray + isotope substitution technique | Glass | Ni–B | [21], | Ni–Nb | [22] |
| (E) X-ray + polarized neutron technique | Glass | Co–P | [12], | Tb–Fe | [23] |
| (F) X-ray anomalous scattering (AXS) | Glass | Ge–Se | [24], | Ni–Zr | [25] |
| | liquid | Ni–Si | [26], | $GeBr_4$ | [27] |
| (G) X-ray anomalous scattering + neutron | Glass | $GeO_2$ | [28, 29, 30] | | |

the isotopes. Thus, the fundamental concept of techniques (i) and (ii) encompasses not only the chemical identity but also the structural identity. This assumption may be valid in a thermodynamically equilibrated disordered state such as in liquids. However, we should keep in mind that such structural identity may not persist in a thermodynamically meta stable glassy state, because some properties of the as-quenched amorphous alloys are known to vary from one production run to another and from one part of the ribbon to another. This clearly suggests that there is some variation in chemical short-range order at the microscopic level.

On the other hand, techniques (iii) and (iv) are free from this ambiguity, because it is possible to vary the weighting factors without the use of different samples. It is also worth mentioning that the anomalous neutron scattering is limited to several isotopes only, such as $^{6}Li$, $^{10}B$, $^{113}Cd$, $^{149}Sm$, $^{157}Eu$ and $^{157}Gd$, although the variation arising from the anomalous dispersion effect of neutrons is several times larger than in the X-ray case.

In anomalous X-ray scattering, the atomic scattering factor should include the anomalous dispersion term of $f'$ and $f''$. The square of the mean scattering factor, $\langle f\rangle^2$, and the mean square average of the scattering factor, $\langle f^2\rangle$, are then expressed by

$$\langle f\rangle^2 = \left[\sum c_i f_i(Q,E)\right]^2 = \langle f\rangle\langle f^*\rangle = \langle (f^\circ + f')\rangle^2 + \langle f''\rangle^2, \tag{3.18}$$

$$\langle f^2\rangle = \sum c_i f_i^2(Q,E) = \langle (f^\circ + f')^2\rangle + \langle (f'')^2\rangle. \tag{3.19}$$

The so-called Laue monotonic scattering term in (2.29) and the weighting factor, $w_{ij}$, in (3.14) can also be written as follows:

$$\left[\langle f^2\rangle - \langle f\rangle^2\right] = c_A\,(1-c_A)f_A f_A^* + c_B(1-c_B)f_B f_B^* \\ - \; 2c_A c_B\left[(f_A^\circ + f_A')(f_B^\circ + f_B') + f_A'' f_B''\right], \tag{3.20}$$

$$\left.\begin{aligned} w_{AA} &= c_A^2\left[(f_A^\circ + f_A')^2 + (f_A'')^2\right]/\langle f\rangle^2 \\ w_{BB} &= c_B^2\left[(f_B^\circ + f_B')^2 + (f_B'')^2\right]/\langle f\rangle^2 \\ w_{AB} &= c_A c_B\left[(f_A^\circ + f_A')^2(f_B^\circ + f_B'')^2 + f_A'' f_B''^2\right]/\langle f\rangle^2 \end{aligned}\right\}. \tag{3.21}$$

The measurements of the X-ray scattering intensity at two energies near the absorption edge of the constituent element provide additional items of the so-called total structure factor of a binary system. (For example, the measurements with Fe-K$\alpha$ and Fe-K$\beta$ radiation in Fig. 3.1.) These data, when coupled with those obtained from the measurement at an energy away from the absorption edge (the measurement with Mo-K$\alpha$ radiation in Fig. 3.1), permit separation of the three partial structure factors in the manner of (3.17).

The partial structure functions provide unique quantitative information for describing the atomic-scale structure of non-crystalline materials. However, as shown in Chap. 1, the actual implementation of the respective partial structure functions from measured structure data is not a trivial task even for a binary system. In other words, the complete separation of partial structure functions is extremely difficult in multi-component, non-crystalline materials containing more than three elements. The AXS method appears to hold promise in reducing this difficulty with sufficient reliability by obtaining the environmental structure function around a specific element as shown in the following.

As shown in Fig. 3.8 using an $A$–$B$–$C$ ternary case as an example, three environmental structure functions can be obtained from the AXS measurements at energies close to the absorption edge of $A$, $B$ and $C$, respectively. In this environmental analysis around a specific element, we are free from difficulties arising from the complete separation of six partial functions for individual constituent pairs. The following merit may also be stressed with respect to the AXS method, especially for liquid samples. A liquid sample is usually contained in a cell with windows transparent to X-rays. For quantitative structural analysis, the intensity from a liquid sample alone is estimated by accurately correcting the scattering intensity from the window materials. This is usually carried out using the intensity only from the window materials corrected for absorption by the liquid sample. On the other hand, in the schematic of Fig. 3.9, each scattering intensity profile measured at two energies, $E_1$ and $E_2$, contains the contribution from the window materials as well as that from the liquid sample [31]. When these two energies are close to the absorption edge of the $A$ element in a sample, by taking the difference

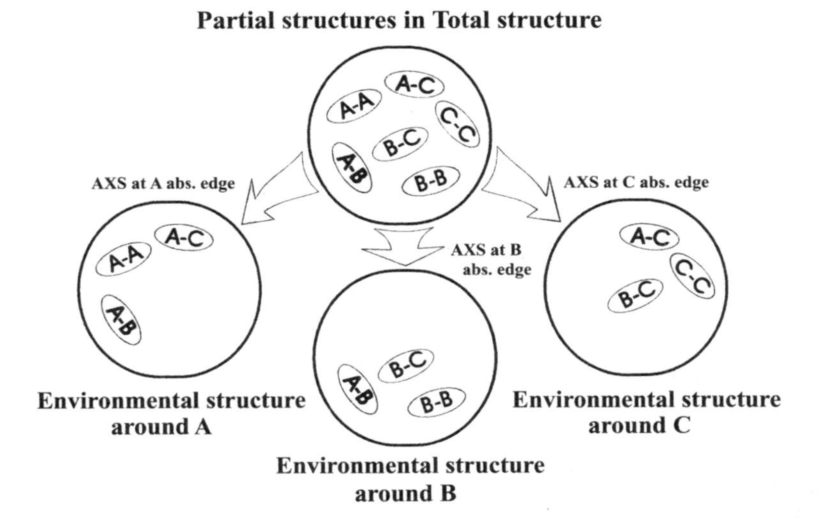


**Fig. 3.8.** Advantage of AXS method for reducing the number of atomic pairs in a $A$–$B$–$C$ ternary system

between the two profiles, the contributions from the window materials, as well as those from non-$A$ components in the present case, are automatically eliminated. In this way, we are relieved of the tedious correction procedure for the window materials. This process was developed by analogy to the AXS measurement in an amorphous thin film grown on a substrate [32]. It is possible to vary the weighting factors without the use of different samples in the AXS method. This also contrasts with the situation of the isotopic substitution measurements requiring different samples in which the structure is automatically assumed to remain identical upon substitution of the isotope.

The environmental structure function analysis is based on the idea that the difference observed between two profiles contains information only about the component that is scattering X-rays anomalously. In other words, the energy derivative of the measured X-ray scattering intensity at a constant wave vector is dominated by the change arising from the energy dependence of the anomalous dispersion factors, $(\mathrm{d}f'/\mathrm{d}E)_Q$ and $(\mathrm{d}f''/\mathrm{d}E)_Q$. For this reason, the angular scanning measurement coupled with the energy-dispersive mode using the energy-sensitive, solid-state detector (often called SSD) gives the X-ray scattering intensity covering a wide energy range in which the AXS is well appreciated. The basic concept of this energy-derivative technique corresponds to the so-called frequency-modulated X-ray diffraction first proposed by Shevchik [33]. When the incident X-ray energy is tuned close to the ab-

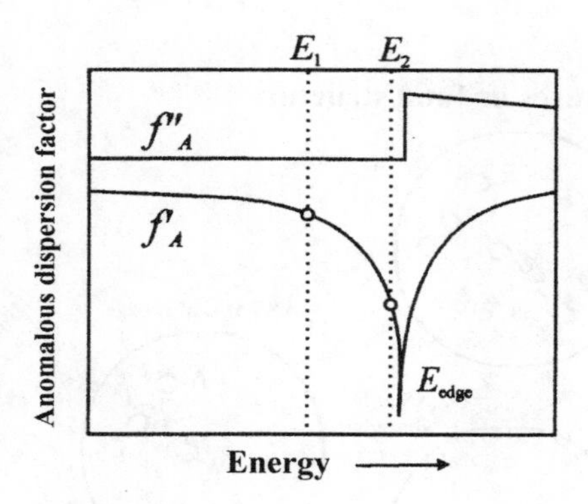


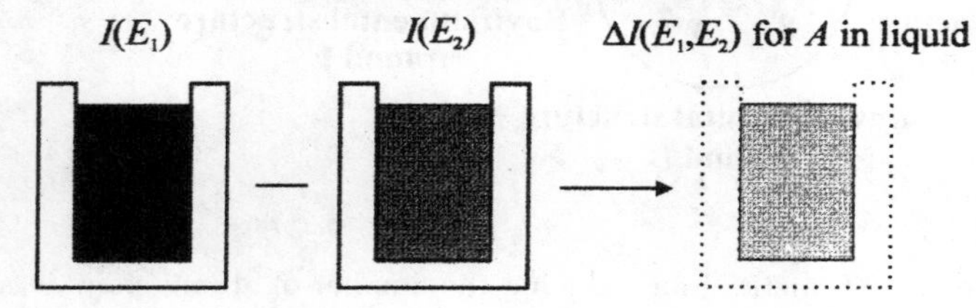


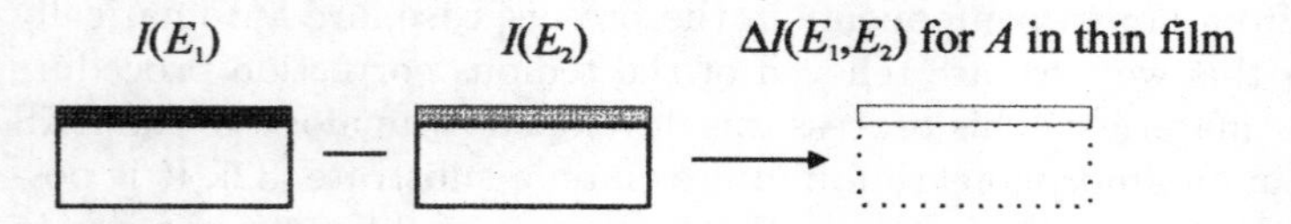


**Fig. 3.9.** Advantage of AXS method for measuring the X-ray scattering intensity of a liquid sample in a cell or a film sample grown on a substrate

sorption edge for one of the elements involved, the energy derivative of the scattered intensity data is strongly affected by the interference from the corresponding element with a sensitivity greater than in the case of the usual AXS measurement by the common characteristic K radiation [34, 35].

These interesting features clearly suggest that use of a strong-intensity X-ray source such as synchrotron radiation can provide valuable structural information for multi-component, non-crystalline materials, using the energy-derivative mode combined with AXS.

## References


1. R.W. James: *The Optical Principles of the Diffraction of X-rays* (G.Bells, London 1954)
2. H. Wagenfeld in: *Anomalous Scattering*, ed. by S. Ramaseshan and S.C. Abraham (Int. Union of Crystallography, Munksgard, 1975) pp. 13
3. D.H. Templeton, L.K. Templeton, J.C. Phillips and K.O. Hodgson: Acta Crystallogr., A **36**, 436 (1980)
4. Y. Waseda: *Novel Application of Anomalous X-ray Scattering for Structural Characterization of Disordered Materials* (Springer, Heidelberg, Berlin, New York 1984)
5. Photon Factory (Tsukuba): available at http://pfwww.kek.jp/
6. M.C. Nichols, D.K. Smith and Q. Johnson: J. Appl. Crystallogr., **18**, 8 (1985)
7. I.G. Wood, L. Nicholls and G. Brown: J. Appl. Crystallogr., **19**, 371 (1986)
8. K. Sugiyama, A. Shikida, Y. Waseda, J. Hino and J.M. Toguri: Shigen-to-Sozai, **106**, 485 (1990)
9. *International Tables for X-ray Crystallography, Vol.II,III and IV* (Reidel, Dordrecht 1985)
10. K. Shinoda, K. Sugiyama, C. Reynales, Y. Waseda and K.T. Jacob: Shigen-to-Sozai, **111**, 801 (1995)
11. Y. Waseda, K. Shinoda and K. Sugiyama: Z. Naturforsch., **50a**, 1199 (1995)
12. J.F. Sadoc and J. Dixmier: *Proc. 2nd Int. Conference on Rapidly Quenched Metals*, Cambridge, Mater. Sci. Eng., **23**, 187 (1976)
13. J.E. Enderby: *Physics of Simple Liquids*, ed. by N.H. Temperley, J.S. Rowlinson, G.S. Rashbrooke (North-Holland, Amsterdam 1968) pp. 612
14. T. Masumoto, T. Fukunaga and K. Suzuki: *Abstract of Topical Conf. on the Atomic Scale Structure of Amophous Solids*, Yorktown Heights, New York(1978), p. F-7; Bull. Am. Phys. Soc., **23**, 467 (1978); Sci. Rep. Res. Inst. Tohoku Univ., Sendai, **28A**, 208 (1980)
15. F. Paasche, H. Olbrich, G. Rainer-Harbach, P. Lamparter and S. Steeb: Z. Naturforsch., **37a**, 1215 (1982)
16. T. Mizoguchi, T. Kudo, T. Irisawa, N. Watanabe, N. Niimura, M. Misawa and K. Suzuki: *Proc. 3rd Int. Conf. on Rapidly Quenched Metals*, Brighton (1978), The Metals Society (London), Cof. Proc. No.198, (1978), pp. 384; J. Phys. Soc. Jpn., **45**, 1773 (1978)
17. K. Yamada, Y. Endoh, Y. Ishikawa and N. Watanabe: J. Phys. Soc. Jpn., **48**, 922 (1980)
18. J.E. Enderby, P.A. Egelstaff and D.M. North: Philos. Mag., **14**, 961 (1966)
19. E.W.J. Mitchell, P.F.J. Ponet and R.J. Stewart: Philos. Mag., **34**, 721 (1976)
20. J. Bletry and J.F. Sadoc: J.Phys. F., **5**, L110 (1975)
21. P. Lamparter, W. Sperl, S. Steeb and J. Bletry: Z. Naturforsch., **37a**, 1223 (1982)
22. E. Sváb, F. Forgacs, F. Hajdu, N. Kroó and J. Takács: J. Non-Cryst. Solids, **45**, 1773 (1978)
23. W.P. O'Leary: J.Phys. F., **5**, L175 (1975)
24. P.H. Fuoss, W.K. Warburton and A. Bienenstock: J.Non-Cryst. Solids, **35/36**, 1223 (1980)
25. J.C. De Lima, J.H. Tonnerre and D. Raoux: J. Non-Cryst. Solids, **106**, 38 (1988)
26. Y. Waseda and S. Tamaki: Philos. Mag., **32**, 951 (1975)
27. K.F. Ludwig Jr., W.K. Warburton and L. Wilson: J. Chem. Phys., **87**, 604 (1987)
28. P. Bondt: Acta Crystallogr., A **30**, 470 (1974)

29. Y. Waseda, K. Sugiyama, E. Matsubara and K. Harada: Mater. Trans. JIM, **32**, 421 (1990)
30. D.L. Price, M.L. Saboungi and A.C. Barnes: Phys. Rev. Lett., **81**, 3207 (1998)
31. E. Matsubara and Y. Waseda: J. Phys. Condens. Matter, **1**, 8575 (1989)
32. Y. Waseda, E. Matsubara, K. Okuda, K. Omote, K. Tohji, S.N. Okuno and K. Inomata: J. Phys. Condens. Matter, **4**, 6355 (1992)
33. N.J. Shevchik: Philos. Mag., **35**, 805 and 1289 (1977)
34. P.H. Fuoss, P. Eisengerger, W.K. Warburton and A. Bienenstock: Phys. Rev. Lett., **46**, 1537 (1981)
35. R.G. Munro: Phys. Rev., B **25**, 5037 (1982)

# 4. Experimental Determination of the Anomalous Dispersion Factors of X-rays – Theoretical and Experimental Issues

Various aspects on the anomalous dispersion effect of X-rays have already been described (see, for example, [1, 2]). The real component of the anomalous dispersion factors, $f'$, is essential for the accurate structural analysis of both crystalline and non-crystalline materials, whereas the imaginary component, $f''$, represents the absorption and it is necessary for the data processing for diffraction or XAFS (X-ray absorption fine structure) measurements. The complex correction factor $(f' + \mathrm{i}f'')$ can also be used for phase determination of the so-called crystallographic structure factor or determination of the partial structure functions in multi-component, non-crystalline materials. This is particularly true when coupled with an intense, white X-ray source that allows substantial progress in the structural study of various materials and various states (see, for example, [3]). Thus, in order for the anomalous X-ray scattering(AXS) method to provide much valuable structural information, we must characterize well the magnitude and energy variation of the anomalous dispersion factors, $f'$ and $f''$.

## 4.1 Theoretical Estimation of the Anomalous Dispersion Factors

The values of the anomalous dispersion factors have been theoretically calculated for six energies corresponding to the frequently used characteristic K radiation of Ag, Mo, Cu, Co, Fe and Cr (see, for example, [4, 5]); they are compiled in the *International Tables for X-ray Crystallography.* It may safely be said that agreement with experiments is recognized, particularly at the lower-energy side of the absorption edge. However, the energy-variable measurement for the AXS method can use a wide range of the energy spectrum for structural determination. This requires sufficiently reliable knowledge of the anomalous dispersion factors as a continuous function of energy. The theoretical aspects on the anomalous dispersion factors of X-rays have been described in detail (see, for example, [6, 7, 8]). Therefore, we give here a brief background on the theoretical aspects of the parameters that are necessary for the efficient use of the AXS method.

Following the description given by Fukamachi [8] and James [9], the fundamental equations for evaluating the anomalous dispersion factors can be expressed as follows:

$$f'(E) = -\frac{1}{2}\int\left(\frac{\mathrm{d}g_{oj}}{\mathrm{d}E_{jo}}\right) \times E_{jo}\left(\frac{E_{jo}-E}{(E_{jo}-E)^2+\gamma_{oj}^2/4}+\frac{E_{jo}+E}{(E_{jo}+E)^2+\gamma_{oj}^2/4}\right)\mathrm{d}E_{jo}, \quad (4.1)$$

$$f''(E) = \frac{1}{2}\int\left(\frac{\mathrm{d}g_{oj}}{\mathrm{d}E_{jo}}\right)E_{jo}\frac{\gamma_{oj}/2}{(E_{jo}-E)^2+\gamma_{oj}^2/4}\mathrm{d}E_{jo}, \quad (4.2)$$

where $E$ corresponds to the photon energy and its subscript denotes the state of photon, i.e. the initial (zero) and the $j$-th scattering process. $\gamma_{oj}$ refers to the convoluted width of states zero and $j$. $g_{oj}$ is the so-called oscillator strength, which is defined by the following equation in the atomic unit ($h = m = e = 1$):

$$\mathrm{g}_{oj} = \left(\frac{2}{E_{jo}}\right)|\langle\phi_{o}|\boldsymbol{e}\cdot\mathbf{P}|\phi_j\rangle|^2. \quad (4.3)$$

Here $\phi_j$ denotes the wave function of atoms in state $j$, $\boldsymbol{e}$ is the unit vector and $\mathbf{P}$ represents the momentum of the electrons. When coupled with the simple approximation $\gamma_{oj}$ approaching zero, (4.1) and (4.2) can easily be reduced to the following equations, similar to the expression given by classical theory where we use the relation $\lim \varepsilon/(x^2+\varepsilon) = \pi\delta(x)$ at $\varepsilon \to 0$ [9]:

$$f'(E) = \int\frac{(\mathrm{d}g_{oj}/\mathrm{d}E_{jo})E_{jo}^2}{E^2-E_{jo}^2}\mathrm{d}E_{jo}, \quad (4.4)$$

$$f''(E) = \pi/2\int E_{jo}\left(\frac{\mathrm{d}g_{oj}}{\mathrm{d}E_{jo}}\right)\delta(E_{jo}-E)\mathrm{d}E_{jo} = \left(\frac{\pi E}{2}\right)\left(\frac{\mathrm{d}g}{\mathrm{d}E}\right). \quad (4.5)$$

Equation (4.5) gives the relation $(\mathrm{d}g_{oj}/\mathrm{d}E_{jo}) = (2/\pi E_{jo}) \times [f''(E_{jo})]$ , when $\gamma_{oj} \to 0$, and hence the so-called *dispersion relation* or *Kramers–Krönig relation* is obtained:

$$f'(E) = -\frac{1}{\pi}\int[f''(E_{jo})]_{\gamma_{oj}\to 0} \times\left(\frac{E_{jo}-E}{(E_{jo}-E)^2+\gamma_{jo}^2/4}+\frac{E_{jo}+E}{(E_{jo}+E)^2+\gamma_{jo}^2/4}\right)\mathrm{d}E_{jo}. \quad (4.6)$$

This can also be expressed by the following simpler form, as already used in (4.2):

$$f'(E) = \frac{2}{\pi}\int_0^\infty\frac{f''(E_{jo})E_{jo}}{E^2-E_{jo}^2}\mathrm{d}E_{jo}. \quad (4.7)$$

The anomalous dispersion factors $f'$ and $f''$ can be estimated from information on the energy variation of the oscillator strength $(\mathrm{d}g/\mathrm{d}E)$. In that regard, the following four methods may be suggested for the major calculations of the anomalous dispersion factors that have been reported to date:

(i) Non-relativistic dipole approximation (NRDP) method (e.g. [10,11,12]).
(ii) Semi-empirical method with a dispersion relation (e.g. [13,14]).
(iii) Relativistic dipole approximation (RDP) method (e.g. [4,5]).
(iv) Relativistic multipole approximation (RMP) method (e.g. [15,16])

Hönl [10] proposed the method for evaluating the oscillator strength in terms of the non-relativistic wave function of a hydrogen atom with some modification for atoms containing more than two electrons and provided the following equation stands for the K-shell electrons:

$$\left(\frac{\mathrm{d}g}{\mathrm{d}E}\right)_{\mathrm{K}} = \frac{2^8 \mathrm{e}^{-4}}{9E_{\mathrm{K}}}\left[\frac{4}{(1-\delta_{\mathrm{K}})^2}\left(\frac{E_{\mathrm{K}}}{E}\right)^3 - \frac{7}{(1-E_{\mathrm{K}})}\left(\frac{E_{\mathrm{K}}}{E}\right)^4\right], \qquad (4.8)$$

where $E_{\mathrm{K}} = E_{\circ}(1-\delta_{\mathrm{K}})$, $E_{\circ}$ is the threshold energy of absorption in the hydrogen-like model and $\delta_{\mathrm{K}}$ corresponds to the parameter containing the higher-order electron effect on the ground-state energy.

By introducing (4.8) into (4.4) and (4.5), one can estimate $f'(E)$ and $f''(E)$. This method has also been used for the L-shell electrons by Eisenlohr and Müller [11] and for the M-shell electrons by Wagenfeld [12], including quadrupole and higher-order terms, which are generally small (about 1% of the dipole term). In the Hönl approach, only transitions to the positive-energy final state are taken into account, and a constant screening parameter is utilized for the inner K-shell electrons. Nevertheless, good agreement between calculation and experiment is found in some cases. However, it should be remembered that the results obtained by this method are no doubt less accurate than those of other methods, particularly for heavy elements–atomic numbers above 30, because of the use of non-relativistic wave functions.

Even in the framework of classical mechanics, it is possible to calculate the anomalous dispersion factors, $f'(E)$ and $f''(E)$, when coupled with the experimental absorption data, because the oscillator strength can be related to the energy dependence of the linear absorption coefficient. This is readily explained by the relationship between the macroscopic refractive index and the linear absorption coefficient of X-rays (see, for example, [17]).

Along the line of this concept, Parratt and Hempstead [13] proposed a simple but useful semi-empirical method for calculating the values of $f'(E)$ and $f''(E)$. The measured absorption data, $\mu(E)$, at energy $E$ can be described by the following simple form:

$$\left.\begin{aligned}\mu(E) &= \mu_{\mathrm{K}}(E_{\mathrm{K}}/E)^{n_{\mathrm{K}}}\,, && E > E_{\mathrm{K}}\\ &= 0\,, && E < E_{\mathrm{K}}\end{aligned}\right\}, \qquad (4.9)$$

where $E_{\mathrm{K}}$ is the energy of the absorption edge, $\mu_{\mathrm{K}}$ is the linear absorption coefficient at energy $E_{\mathrm{K}}$ and $n_{\mathrm{K}}$ corresponds to the polynominal index. The parameters $\mu_{\mathrm{K}}$ and $n_{\mathrm{K}}$ can be found by fitting a curve of (4.9) to measured absorption data. Since the energy derivative of the oscillator strength $(\mathrm{d}g/\mathrm{d}E)$ is proportional to $\mu(E)$, the following form, similar to (4.9), may also be used:

$$(\mathrm{d}g/\mathrm{d}E)_{\mathrm{K}} = g'_{\mathrm{K}}(E_{\mathrm{K}}/E)^{p_{\mathrm{K}}}, \qquad E > E_{\mathrm{K}}. \qquad (4.10)$$

When the quantity $g_{\mathrm{K}}$ is defined as

$$g_{\mathrm{K}} = \int_{E_{\mathrm{K}}}^{\infty} (\mathrm{d}g/\mathrm{d}E)\mathrm{d}E, \tag{4.11}$$

the following simple relation can readily be obtained:

$$g'_{\mathrm{K}} = (P_{\mathrm{K}} - 1)\frac{g_{\mathrm{K}}}{E_{\mathrm{K}}}. \tag{4.12}$$

The energy derivative of the oscillator strength can then be expressed by

$$\left(\frac{\mathrm{d}g}{\mathrm{d}E}\right)_{\mathrm{K}} = \frac{p_{\mathrm{K}} - 1}{E_{\mathrm{K}}} g_{\mathrm{K}} \left(\frac{E_{\mathrm{K}}}{E}\right)^{p_{\mathrm{K}}} = \frac{c^{\circ}}{2\pi^2}\mu_{\mathrm{K}} \left(\frac{E_{\mathrm{K}}}{E}\right)^{p_{\mathrm{K}}}, \quad E > E_{\mathrm{K}}, \tag{4.13}$$

where $c^{\circ}$ is the speed of light.

On the basis of this idea, Parratt and Hempstead [13] subtracted the fitting part to the experimental data ($g_{\mathrm{K}}$ or $\mu_{\mathrm{K}}$) and numerically integrated the residue in a way similar to the case of (4.4) and (4.5); they then developed equations for evaluating the anomalous dispersion factors with an appropriate value of $p_{\mathrm{K}}$. Following their procedure, Dauben and Templeton [14] calculated the anomalous dispersion factors for three common characteristic radiations (Cr-K$\alpha$, Cu-K$\alpha$, and Mo-K$\alpha$) by using the measured linear absorption coefficient, $\mu(E)$. The results were also compiled in the *International Tables for X-ray Crystallography.* The values calculated by Dauben and Templeton [14] include the angular variation of $f'$ and $f''$ by multiplying the dispersion correction of each shell. However, it should be noted that this angular dependence has no physical significance theoretically, because the semi-empirical method accounts only for the dipole approximation, which has no angular dependence (see, for example, [12]).

Parratt and Hempstead [13] proposed the following values: $p_{1\mathrm{s}} = 11/4 = 2.75$, $p_{2\mathrm{s}} = 7/3 = 2.33$ and $p_{\mathrm{others}} = 5/2 = 2.5$. However, the assumption that the K-shell electrons possess the same value (2.75) for all cases is no doubt a crude approximation. It should also be noted that Weiss [18] shows that a change in $p_{\mathrm{K}} = 2.7 \pm 0.1$ produced variation of the order of $\pm 10\%$ in the resulting $f'(E)$. Cromer [19] also estimated the anomalous dispersion factors for several characteristic radiations using the idea proposed by Parratt and Hempstead [13]. In his case, the oscillator strength, $g_{\mathrm{K}}$, in (4.13) was theoretically calculated from the wave function of the Hartree type with the Thomas–Reich–Kuhn sum rule. The physical significance of this semi-empirical method depends mainly on the accuracy of the parameter $p_{\mathrm{K}}$ (or $n_{\mathrm{K}}$) in reproducing the linear absorption coefficient obtained experimentally.

On the other hand, an attempt by Cromer and Liberman [4,5] was made to theoretically evaluate the anomalous dispersion factors for most of the elements ($Z = 3$–$98$) by using the relativistic wave function. Their expressions are of considerable length and somewhat complicated, but the basic concept of their method refers to the theoretical evaluation of the oscillator strength by using the more realistic wave function of atoms in the respective state.

The anomalous dispersion factors are then calculated in the same manner as the other two methods employing (4.4) and (4.5). However, in contrast to the semi-empirical method, the integration for these equations was made numerically without approximation to the form of the cross-section vs energy curve expressed by (4.9) or (4.10). Cromer and Liberman [4] used the relativistic wave function using a Kohn–Sham model for exchange [20] with Bearden's experimental energy levels [21] and the procedure for evaluating the relativistic photoelectron absorption coefficient given by Brysk and Zerby [22] and provided the anomalous dispersion factors of many elements for five characteristic K radiations. It is rather stressed here that their results for both $f'$ and $f''$ show reasonable agreement with the experimental data as shown in Table 4.1. This table includes the results of Gerward et al. [23].

In the RDP method employed by Cromer and Liberman [4], the relativistic correlation term, corresponding to the forward scattering amplitude at the higher energy limit is described as follows:

**Table 4.1.** Numerical examples for comparison of experimental and calculated anomalous dispersion factors using the results for Ge and Si

| | Radiation | Element | Calculation | Experiment |
|---|---|---|---|---|
| $f'$ | Cu-K$\alpha$ | Ge | –1.803 | –1.79 |
| | Mo-K$\alpha$ | | 0.082 | 0.06 |
| | Ag-K$\alpha$ | | 1.142 | 0.27 |
| $f''$ | Cu-K$\alpha$ | Ge | 0.0 | 0.0 |
| | Mo-K$\alpha$ | | 1.588 | 1.58 |
| | Ag-K$\alpha$ | | 1.054 | 1.06 |
| $f'$ | Cr-K$\alpha$ | Si | 0.381 | 0.389 |
| | Fe-K$\alpha$ | | 0.337 | 0.344 |
| | Co-K$\alpha$ | | 0.313 | 0.320 |
| | Cu-K$\alpha$ | | 0.270 | 0.274 |
| | Mo-K$\alpha$ | | 0.098 | 0.099 |
| | Ag-K$\alpha$ | | 0.068 | 0.070 |
| $f''$ | Cr-K$\alpha$ | Si | 0.693 | 0.696 |
| | Fe-K$\alpha$ | | 0.509 | 0.510 |
| | Co-K$\alpha$ | | 0.438 | 0.440 |
| | Cu-K$\alpha$ | | 0.330 | 0.331 |
| | Mo-K$\alpha$ | | 0.071 | 0.071 |
| | Ag-K$\alpha$ | | 0.043 | 0.044 |

Ge: D.T.Cromer and D.Liberman: [4]
Si: L.Gerward, G.Theusen, M.S.Jensen and I.Alstrup: [23]

$$f'_{\mathrm{corr}} = \frac{5}{3} \cdot \frac{E_{\mathrm{tot}}}{mc_{\circ}^{2}}. \tag{4.14}$$

It is also noted that the total kinetic energy, $E_{\mathrm{tot}}$, of the electrons in the atom is, of course, given by a constant value for the particular atom. As shown by Stibius-Jensen [24], the RDP method induces an error in this term compared with the RMP method, although there is no significant difference between the photoelectron absorption coefficients calculated by these different methods. In the RMP method employed by Creagh and McAuley [15], the relativistic correlation term is given by $E_{\mathrm{tot}}/mc_{\circ}^{2}$, and their integration grid includes 24 photon energies in the range of 4.509 to 24.94 keV. In contrast, the photoelectron absorption coefficient for each atom is calculated for ten photon energies in the RDP method of Cromer and Liberman [4].

More recent developments with the detailed relativistic **S**-matrix approach (see, for example, [25,26]) are considered to be much more rigorous than the earlier approaches. In the **S**-matrix approach, the model involves Rayleigh scattering by considering a second-order single transition relevant to electrons bounded in a relativistic self-consistent central potential of the Dirac–Hartree–Fock–Slater type. Changes in polarization on scattering are taken into account (see, for example, [7]). Although corrections to the data sets of Cromer and Liberman are clearly suggested [27], we still have rather sparse results obtained directly from the **S**-matrix calculation because of the requirement of a long computing time, so that no convenient tabular form is available. Only the results for Si and Al have been obtained by the **S**-matrix calculation [28,29]. The **S**-matrix results for Si [28] are $f' = 0.084$ for Mo-K$\alpha$ and $f' = 0.055$ for Ag-K$\alpha$. The RMP method for Si [29] gives the values of $f' = 0.082$ and 0.052, respectively. For most cases, good agreement (within $\pm 3\%$) between the predictions of the RMP method [28] and the available experimental data is found in the energy range away from the absorption edge. Sufficiently reasonable agreement is also recognized between the RMP calculation and the **S**-matrix case [29]. Nevertheless, the anomalous dispersion factors estimated by the Cromer–Liberman scheme still provide a useful guide for selecting the experimental conditions for AXS measurements near the absorption edge for various elements and energies of general interest. With respect to this, the energy variations of the anomalous dispersion factors including the mass-absorption coefficient for 96 elements are easily obtained from the public database of SCM-AXS (http://www.iamp.tohoku.ac.jp/; http://www.tagen.tohoku.ac.jp/), which covers the energy region between 1 and 50 keV. These results are also useful to the energy-dispersive X-ray diffraction (EDXD) method for various materials in a variety of states. For the convenience of future investigations using the AXS method, the energies of absorption edges in keV units for various elements are summarized in Chap. 10. It should also be kept in mind that the ionization state of an atom affects the position of the absorption edge [31,32]. This is particularly true

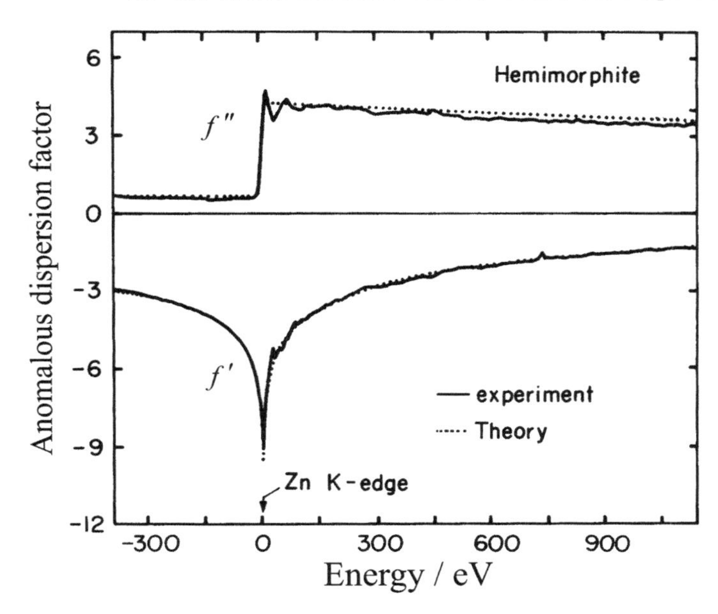


**Fig. 4.1.** Anomalous dispersion factors of Zn in hemimorphite near the K absorption edge. The values of $f'(E)$ are estimated from the absorption data of $f''(E)$ through the dispersion relation [30]

at energy levels close to the absorption edge, although such an energy shift is usually not so distinct.

Theoretical calculations regarding the anomalous dispersion factors, $f'(E)$ and $f''(E)$, described here are restricted to free (isolated) atoms. Hence, none of them includes the near-edge phenomena such as XANES (X-ray absorption near edge structure) and EXAFS (Extended X-ray absorption fine structure) observed at the threshold of the absorption edge. Such characteristic fine structures appear significantly in the narrow energy region, particularly on the higher-energy side of the absorption edge, as exemplified by Fig. 4.1 using the Zn case as an example [30]. In this regard, the values theoretically predicted do not reproduce any fine structure in $f'(E)$ and $f''(E)$ associated with the near-edge phenomena. For this purpose, alternative rigorous consideration of effects such as the band-structure effect and the chemical environmental correlation around atoms scattering X-rays anomalously should be taken into account for the wave functions required for evaluating the oscillator strength (see (4.3)). In the near-edge region, the wave functions are known to vary considerably from one state to the other (see, for example, [9]). Thus, further calculation for the anomalous dispersion factors, particularly in the energy region just beyond the absorption edge, is required to predict the near-edge phenomena. This should be one of the most important future subjects in AXS or in near-edge phenomena.

## 4.2 Experimental Determination of the Anomalous Dispersion Factors

Use of the EDXD technique with a continuous spectrum enables us to treat an energy region where the near-edge phenomenon is appreciable. Synchrotron radiation as a light source provides an extremely high-intensity, easily tunable, and highly monochromatic X-ray source. These recent major technical advances allow the experiments to be performed at an energy where the AXS is the greatest. This also has induced rapidly growing interest in the measurement of anomalous dispersion factors for the respective materials, because one can no longer rely on the theoretical values of the anomalous dispersion factors estimated for an isolated atom in such a narrow energy region in which the near-edge phenomena are clearly observed. The experimental verification and our understanding of the near-edge phenomena are still far from complete. This also includes the detailed information about the anomalous dispersion factors on the higher-energy side (i.e. about 50–100 eV) of the absorption edge, as frequently described by XANES. For this reason, a comparison between the theoretical values and the experimental data focuses on the real part of the anomalous dispersion factor, $f'(E)$. This is fully acceptable because the imaginary part, $f''(E)$, is proportional to the linear absorption coefficient.

Various methods have been attempted to determine the anomalous dispersion factors of X-rays. The respective techniques have their own advantages and disadvantages (see, for example, [29, 33]). Although there is no definite conclusion regarding the superiority of one method over the others at the present time, some methods indicate good feasibility, in parallel with recent progress using the energy-dispersive technique with a synchrotron-radiation source. It is believed that the importance of this subject will increase appreciably in the next five to ten years. For this reason, we provide here the background for the experimental determination of the anomalous dispersion factors with some selected examples.

The methods for determining the anomalous dispersion factors, $f'(E)$ and $f''(E)$, frequently used in the literature can be classified into one of the following four categories, although there are differences in detail: X-ray interferometry, refraction and reflection, and Kramers–Krönig (dispersion relation) transformation of the linear absorption coefficient from diffracted intensities and from Pendellösung fringes in perfect crystals.

### 4.2.1 Interferometry Method

One of the typical optical methods is the use of X-ray interferometry (see, for example, [34, 35]). X-ray interferometry is probably the most accurate method for $f'(E)$ measurements, but in general it has only been applied to a restricted energy region. In this method, we measure the refractive index, $n$, of matter for X-rays, actually its deviation from unity represented by $\delta$:

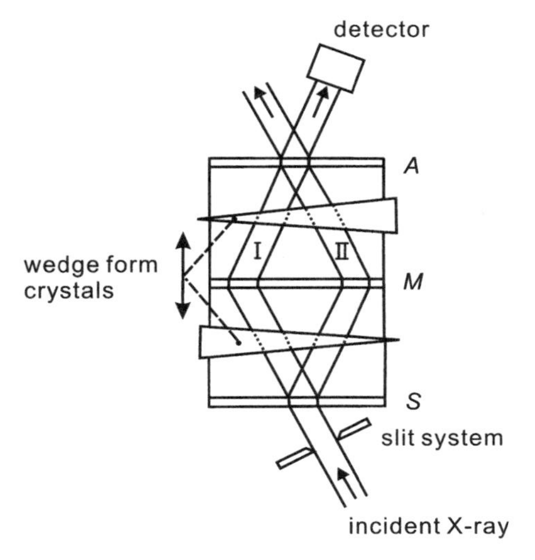


**Fig. 4.2.** Schematic of fundamental relations for X-ray interferometer crystals [36]

$$1 - n = \delta = \alpha + \mathrm{i}\beta = \frac{r_\mathrm{e}\lambda^2}{2\pi}\sum N_j(Z_j + f' + if''), \tag{4.15}$$

where $r_\mathrm{e}$ is the classical electron radius and $\lambda$ is the wavelength of the radiation. $N_j$ and $Z_j$ are the number of atoms per unit volume and the number of electrons (atomic number) of atom $j$, respectively. The value of $\delta$ is usually of the order of $10^{-6}$. Fundamental relations among the interferometer crystals are illustrated in Fig. 4.2 using the work of Bonse and Hellkötter [36] as an example. When one pair of wedge-type crystals are reciprocally shifted along the direction of the arrow, a very small change in fringe pattern can be effectively detected by producing the variation of the optical pass length marked by I and II. Creagh and Hart [37] improved this X-ray interferometer so as to measure the displacement of fringe patterns with and without a foil sample in the beam by changing the basic arrangement with a fixed wedge shown in Fig. 4.3. In this case the sample causes a phase shift and reduction in amplitude. Further improvements have been made to generate fringes in the interferometer and to measure the phase shift, such as the use of phase shifters of parallel-faced samples rotated in the beam or moving the analyzer crystal in the direction normal to the lattice plane (the so-called Angstrom ruler), where a shift by one lattice spacing corresponds to a $2\pi$ phase shift of the fringes [38]. The experimental uncertainty in obtaining $f'(E)$ data is on the order of $\pm 0.1$ electrons, except for in the energy region near and above the absorption edge. In addition, this method requires thin samples, so there are difficulties with brittle materials.

Recently, Bonse and Meterlik [35] constructed the X-ray interferometer device shown in Fig. 4.4. They revealed the very detailed near-edge phenomena in crystalline Ni or Cu foil, with the help of a synchrotron-radiation

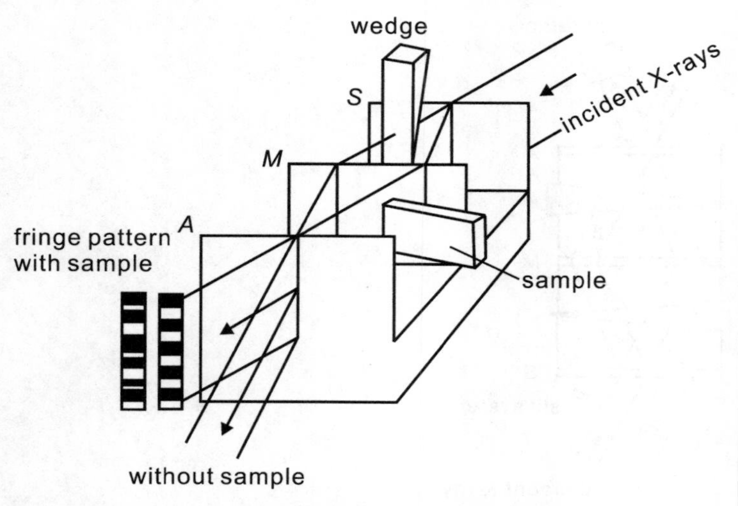


**Fig. 4.3.** Schematic of another X-ray interferometer [37]

source at the Deutschen Elektronen-Synchrotron DESY (Hamburg, Germany). Their results for crystalline Cu foil are given in Fig. 4.5 [39]. The absorption curve related to the $f''(E)$ component was also measured by placing a Cu foil in front of the detector with the phase shifter plane removed from the interfering X-ray beam. Four peaks were found in the absorption curve just above the edge. Such fine structure has frequently been referred to as XANES, which principally differs from the well-known EXAFS (see, for example, [32]). This fine structural change contrasts with the information on the lower-energy side of the absorption edge, where only monotonic energy variation is observed. Similar energy dependence has been well recognized for various materials in the energy region near and above the absorption edge [40, 41], although a direct comparison is not allowed due to the difference in energy resolution and experimental uncertainty. Further experiments for various materials are necessary for obtaining important information regarding the near-edge phenomena.

### 4.2.2 Refraction and Reflection Method

Another typical optical method is the measurement for the deviation from unity in the refractive index through a prism in a manner similar to that conducted in the visible (light) region (see, for example, [42]). Total reflection of X-rays from flat surfaces (see, for example, [43]) is also included in this category. This reflectivity method is also a convenient way for determining the number density of thin layers grown on a substrate by changing the energy of incident X-rays. All these optical methods determine the real component of the anomalous dispersion factors, $f'(E)$, as a phase shift, be-

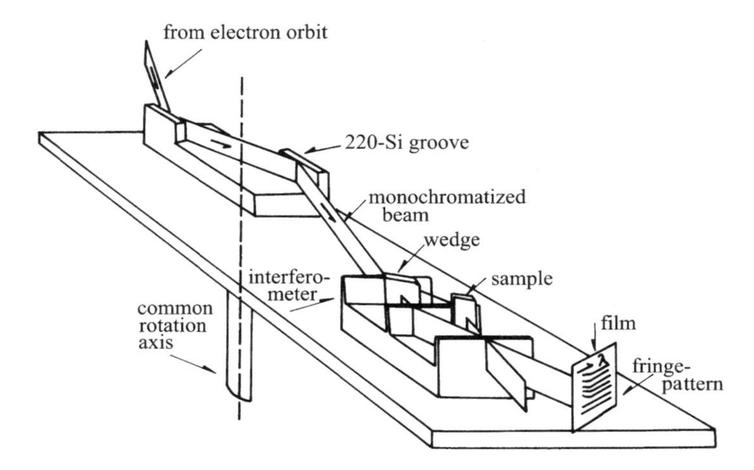


**Fig. 4.4.** Schematic of the X-ray interferometer for synchrotron radiation [35]

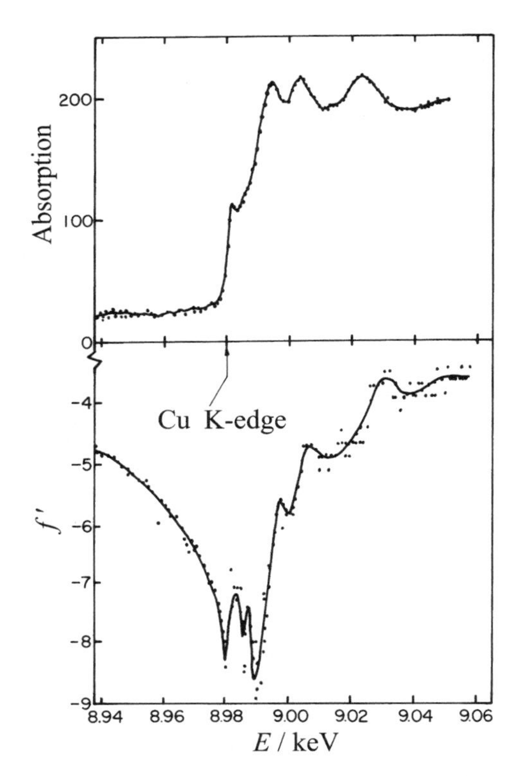


**Fig. 4.5.** Anomalous dispersion factor $f'$ (*bottom*) and absorption (*top*) of Cu foil measured in the energy region near the K absorption edge [39]

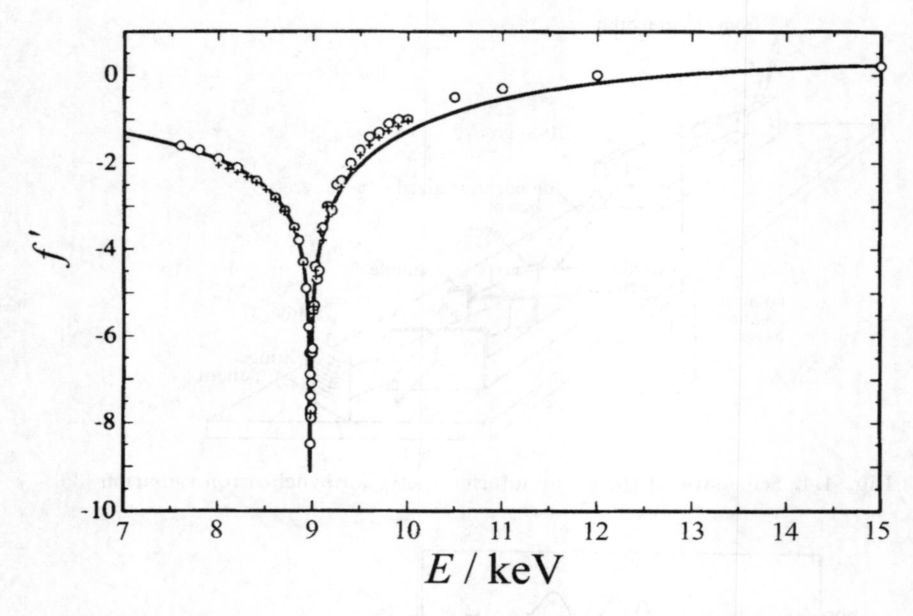


**Fig. 4.6.** Comparison of the $f'$ value for Cu determined by interferometry (+) [44] with those from reflection (○) [43], together with the data calculated by Cromer and Liberman [4]

cause the absorption of sample materials related to the imaginary component, $f''(E)$, gives only the reduction of X-ray intensity in these measurements. For example, the absorption broadens the cutoff in the total reflection or gives no change in the position of the interference patterns. In general, the values obtained by X-ray reflection [43] agree reasonably well with those estimated from X-ray interferometry [44] as exemplified by the result for Cu in Fig. 4.6, although the fine structure should not be ignored in the very close vicinity of the absorption edge (see Fig. 4.5).

### 4.2.3 Intensity Measurement Method

It should be mentioned here that the agreement of measured data with values calculated by Cromer and Liberman's scheme is satisfactory when considering the relativistic correction [27], especially below edge. On the other hand, most of the intensity measurement methods, such as the diffracted intensity measurement for the Friedel pair reflections (see, for example, [45]) and the absorption measurements using two crystals or one crystal with a solid-state detector system (see, for example, [46,47]), are always affected by the absorption of sample materials. Thus, the data processing for extracting $f'(E)$ and $f''(E)$ from measured intensity data is somewhat complicated, in comparison with the cases for the optical method. However, when the absorption measurement can be made with a sufficiently reasonable degree of accuracy, it provides a relatively simple method for determining both anomalous dispersion components, extracting $f'(E)$ and $f''(E)$, by applying the Kramers–Krönig (dispersion) relation. It is stressed that the theoretical basis of this indirect method has been discussed in detail by Kawamura and Fukamachi [48], and its usefulness is now conceptually well recognized and widely used [3,33]. The experimental uncertainty in the absorption experiment of X-rays has been extremely reduced recently, in parallel with the significant technical progress in both detecting systems and X-ray sources, such as the solid-state detector and synchrotron radiation. Therefore, this method appears to be very convenient for determining the anomalous dispersion factors of the particular sample under investigation. For this reason, it may be helpful to recall the essential points of this indirect method.

The imaginary component of the anomalous scattering of X-rays, $f''$, at energy $E$ is directly related to the linear absorption coefficient, $\mu(E)$, through the following equation:

$$f''(E) = \frac{mc_{\mathrm{o}}M}{4\pi N_{\mathrm{A}}^{2} e\rho}\mu(E), \tag{4.16}$$

where $m$ and $e$ are the electron mass and charge, respectively. $M$ is the atomic weight, $N_{\mathrm{A}}$ is the Avogadro's number and $\rho$ is the sample density. The $f''(E)$ curve is a function of energy that corresponds to the so-called EXAFS curve, whichs has received much attention in the determination of the fine structure of both crystalline and non-crystalline materials in a variety of states. In this regard, the EXAFS is a part of the AXS. The real component of extracting $f'(E)$ is given by the dispersion relation as proposed by Kawamura and Fukamachi [48]. The basic equation is (4.7).

It is readily seen from (4.7) that the experimental uncertainty of the real component, $f'(E)$, is directly attributed to that of the imaginary component, $f''(E)$; thus, accurate measurement of the absorption of X-rays should be made for this method. The homogeneity of a sample and the energy resolution of a monochromatic X-ray source are also known to contribute significantly to the accuracy of the absorption measurement, so that the experimental set up should be chosen to optimize these factors. In (4.7), the range of the integral is from zero to infinity, whereas the available experimental data of the energy dependence of the $f''(E)$ curve are usually limited due to technical reasons. That is, measurements in the whole energy range are not possible, and the absorption data including the EXAFS and information of chemical shifts in the close vicinity of the edges are available only in a limited energy region. Also, the contribution of the $f''(E)$ values away from the desired energy is known to be considerably small because of the factor $1/(E^2 - E'^2)$ in (4.7). Therefore, it is quite helpful that the theoretical $\mu(E)$ values are reasonably linked with measured data for the calculation of $f'(E)$ by applying the dispersion relation.

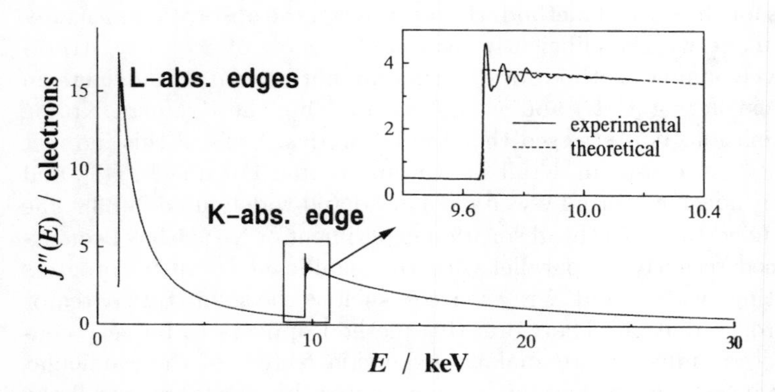


**Fig. 4.7.** The full profile of $f''(E)$ for the calculation of $f'(E)$ for zinc ferrite powder [49]. The *dotted lines* denote the calculated values [50]

With respect to such data processing, the full profile of $f''(E)$ for the calculation of $f'(E)$ is illustrated in Fig. 4.7 using the results for Zn in zinc ferrite powder as an example [49]. The resultant $f'(E)$ and $f''(E)$ values for Zn near the Zn K-absorption edge are given in Fig. 4.8. The anomalous dispersion factors calculated by the Cromer–Liberman scheme [50] are also given in this figure for comparison. Of course, the integration limit of (4.7) has, more or less, influence on the resultant magnitude of the $f'(E)$ values. Such behavior can be seen in the results of Fig. 4.8 estimated from the $f''(E)$ data at three different energy ranges. Namely, when only very limited $f''(E)$ data are used in the integration, one finds some different features such as a positive shift of 0.5 electrons from the theoretical values, although the essential profile itself is unchanged. This is consistent with observations by Kawamura and Fukamachi [48].

The agreement between the experimental data of $f'(E)$ and $f''(E)$ and the calculated values of the Cromer–Liberman scheme is, in the author's view, surprisingly good, because the characteristic behavior of the theoretical anomalous dispersion factors includes a relatively narrow and deep minimum in $f'(E)$ or a sharp increase in $f''(E)$ near the absorption edge. However, the free-atom assumption of the calculated values limits their applicability to structural characterization using the AXS method, mainly because of the neglect of solid-state effects, such as white lines, which result from transitions to a high density of unoccupied states [51,52]. In other words, the solid-state effects are considered to be less important at the lower-energy side of the absorption edge, and the calculated anomalous dispersion factors may be reasonably used for structural analysis. This is one of the main reasons "why we should choose the lower-energy side of the absorption edge for the AXS measurements."

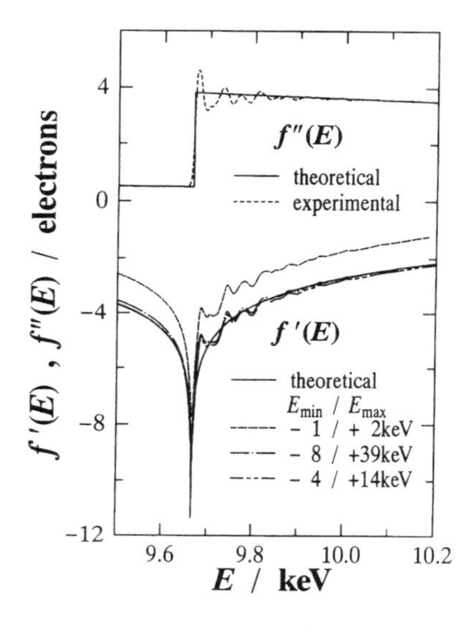


**Fig. 4.8.** Energy variation of the anomalous dispersion factors for Zn in zinc ferrite powder [49]. The *solid lines* denote the calculated values [50]

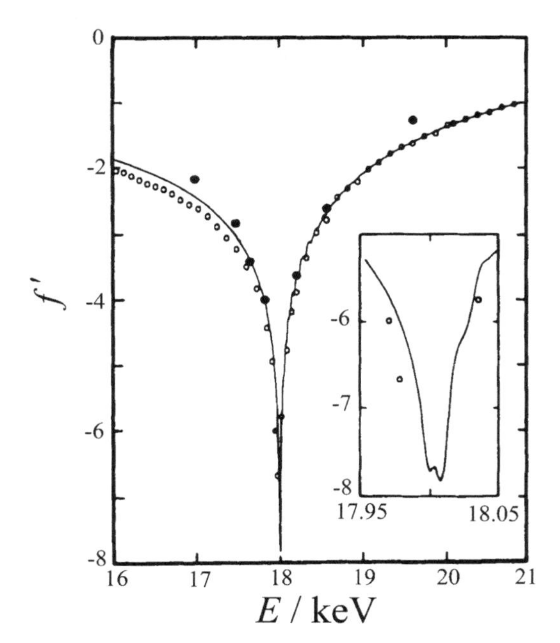


**Fig. 4.9.** Comparison of the $f'$ value of Zr determined by interferometry (*open circles* [53] and *solid* [54]) with those of absorption measurements with the dispersion relation (*solid line* [55])

Accurate determination of the absolute values for the anomalous dispersion factors, $f'$ and $f''$, is one of the difficult experiments. On the other hand, relative value determination, such as the energy dependence of $f'(E)$ and $f''(E)$, may be easier to carry out for both crystalline and non-crystalline systems. The best strategy may be to match the $f'(E)$ data obtained from the intensity measurements coupled with the dispersion relation at certain points to the values independently determined by the optical method. For example, Fig. 4.9 shows a comparison of the experimental data for Zr obtained by interferometry [53, 54] with those of the intensity measurement method with the dispersion relation [55]. It should also be mentioned that a constant value of +0.138 electrons has to be added to the intensity measurement data, as already discussed by Lengler [33], as a relativistic correction [27]. The overall agreement is satisfactory, although the difference of 0.2 electrons should be noted. A similar conclusion is obtained for many metallic elements such as Ti, Ni, Cu, etc. (see for example Hoyt et al [40]). Thus, this method appears to be a relatively easy way to obtain the real part of the anomalous dispersion factors of a desired sample experimentally and is valuable in cases where sample preparation for the optical method is found to be technically difficult. Different chemical binding states, such as Ni in the metal and in the oxide, might be cases in point (see, for example, [43, 56]).

The intensity of X-rays diffracted by crystals is known to depend upon the so-called crystallographic structure factors [9]. Therefore, a number of data sets of the integrated intensities collected with several different energies (wavelengths) can be used for determining the real part of the anomalous dispersion factors as well as the crystal-structure parameters by the least-squares method when a sample structure is known. Suortti et al. [57] applied this method to Ni powder based on measurements at nine energies below the Ni-K edge. Will et al. [58] measured the integrated intensities of a $Yb_2O_3$ powder sample with four energies slightly lower than the Yb $L_{III}$ absorption edge using synchrotron radiation and then analyzed the results by applying a powder least-squares (POWLS) program for structural refinement and determination of the $f'$ values. An experimental uncertainty of $\pm 0.2$–$0.5$ electrons is suggested with this method.

When a sample crystal shows a non-centrosymmetric nature, another method may be utilized for determining the $f'$ values from the relative intensities of certain Bragg peaks denoted by Friedel pairs (see, for example, [45, 59]). In this case, the intensity ratio $I_H/I_{\bar{H}}$ is essential information. When the contribution from the temperature factor is excluded for simplification, the intensity ratio is described in the following form using the case of (111) reflection of the GaP perfect crystal [60]:

$$\frac{I_H}{I_{\bar{H}}} = |F_{111}/F_{\bar{1}\bar{1}\bar{1}}|^2 = \frac{(f^{\circ}_{\mathrm{Ga}} + f'_{\mathrm{Ga}} + f''_{\mathrm{P}})^2 + (f^{\circ}_{\mathrm{P}} + f'_{\mathrm{P}} - f''_{\mathrm{Ga}})^2}{(f^{\circ}_{\mathrm{Ga}} + f'_{\mathrm{Ga}} - f''_{\mathrm{P}})^2 + (f^{\circ}_{\mathrm{P}} + f'_{\mathrm{P}} + f''_{\mathrm{Ga}})^2} . \tag{4.17}$$

In the close vicinity of the Ga–K edge, the sign of the absorption term clearly changes when substituting $(\bar{1}\bar{1}\bar{1})$ for (111).

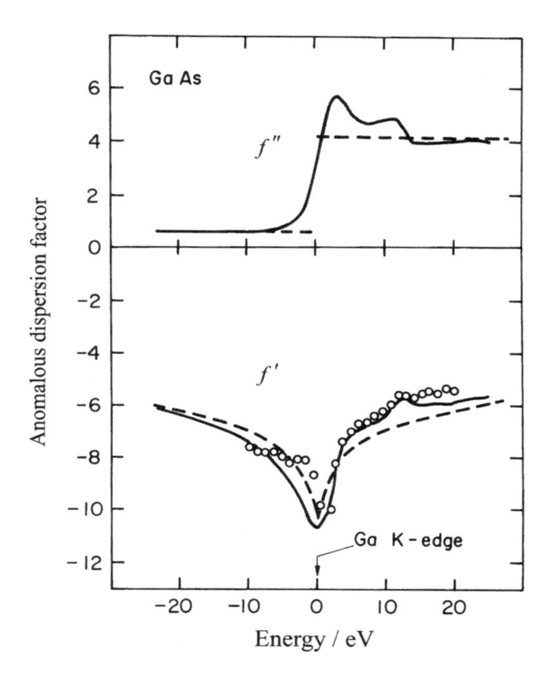


**Fig. 4.10.** Anomalous dispersion factors of Ga in a GaAs crystal near the K absorption edge of Ga. Intensity measurement for the Friedel pairs are shown (*open circles*). The *solid line* for $f'(E)$ is estimated from the $f''(E)$ data through the dispersion relation [61], and the *broken lines* are the calculated values from Hönl theory [10]

Figure 4.10 gives the anomalous dispersion factors of Ga in crystalline GaAs determined by measuring the intensity for the Friedel pair reflections using the results of Fukamachi et al. [61] as an example. In this method, a number of systematic errors encountered in intensity measurements are eliminated by taking the intensity ratio, and the results for some crystals of GaP, GaAs, CdSe, ZnP and ZnS are reported. However, this method is limited to polar perfect crystals only.

Although high-quality single crystals are required and therefore only a few cases, such as Si, Ge and quartz, have been reported, there is another method for determining the $f'$ values from the intensity measurements using the so-called Pendellösing fringes. For high-quality single crystals, the Pendellösing fringes are produced by the interaction between two wave fields with slightly different wave vectors (the transmitted and Bragg reflected wave at a Bragg reflection). In that case, the fringe spacing is described as a function of the crystallographic structure factor (see, for example, [62]). Then, the real part of the anomalous dispersion factors can be estimated from the fringe spacing detected in the Pendellösing fringe pattern when changing the sample thickness (in a wedge-form sample or by inclining a sample crystal) or the X-ray energy. The results of silicon analysis showed that there was fair agreement between the $f'$ values obtained by this method and those of other methods, although only for a very limited number of energies.

In general, the anomalous dispersion factors theoretically calculated by the Cromer–Liberman scheme [4, 5] for an isolated atom agree rather well with the experimental values determined by several methods. In this regard, for most elements the agreement between theory and experiment is within $\pm 3\%$ in the energy range from about 100 eV below to 500 eV above the absorption edge. This is particularly true when the measurement is corrected for overestimation of the relativistic effect in the calculation, as proposed by Kissel and Pratt [27]. This is also a reason why "we preferentially select the lower-energy side of the absorption edge for the AXS measurements."

The following point is also worthy of note. The anomalous dispersion factors of X-rays, in principle, depend not only on the energy but also on the scattering angle. However, the K or L electrons that give the largest contribution to the $f'$ and $f''$ values are distributed in the very narrow region near the atomic nucleus. Therefore, spherically symmetric scattering of X-rays will dominate. Also, no clear evidence has been reported with respect to the angular dependence of the anomalous dispersion factors. For this reason, it may be safely said that the angular dependence of the $f'$ values can be neglected in the AXS method.

Considering the many factors described in this chapter, the $f'(E)$ and $f''(E)$ values compiled in the public database (see, for example, the *International Tables for X-ray Crystallography*) can be used in data processing for analyzing the structure of various materials of interest using the AXS method. The lower-energy side of the absorption edge is likely to fall into this category, because the physical significance of the so-called near-edge phenomena is not severe in that energy region. Furthermore, the relative change in $f'$ as a function of energy is found to be about an order of magnitude easier to obtain, in comparison with the direct application of the anomalous dispersion effect on the absolute scale.

## References


1. S. Ramaseshan and S.C. Abraham (editors): *Anomalous Scattering* (Inter. Union of Crystllogr., Munksgaard, 1975)
2. S. Hosoya: J. Crystallogr. Soc. Jpn., **19**, 68 (1977)
3. G. Metrlik, C.J. Sparks and K. Fischer (editors): *Resonant Anomalous X-ray Scattering* (North-Holland, Amsterdam 1994)
4. D.T. Cromer and D. Liberman: J.Chem. Phys., **53**, 1891 (1970)
5. D.T. Cromer and D. Liberman: Acta Crystallogr., A **37**, 267 (1981)
6. M. Gavrila in: *Inner-shell Processes*, ed. by B. Craseman (Plenum Press, New York 1981) pp. 357

7. L. Kissel and R.H. Pratt in: *Atomic Inner-shell Physics*, ed. by B.Craseman (Plenum Press, New York 1985) Chap. 9
8. T. Fukamachi: J. Crystallogr. Soc. Jpn., **19**, 51 (1977)
9. R.W. James: *The Optical Principles of the Diffraction of X-rays* (G.Bells, London 1954)
10. H.Hönl: Ann. Phys. **18**, 625 (1933)
11. H.Eisenlohr and L.J.Müller: Z. Phys., **136**, 491 and 511 (1954)
12. H. Wagenfeld : Phys. Rev., **144**, 216 (1966)
13. L.G. Parratt and C.F. Hempstead: Phys. Rev. Lett., **94**, 1593 (1954)
14. C.H. Dauben and D. Templeton: Acta Crystallogr., **8**, 841 (1955)
15. D.C. Creagh and W.J. McAuley in: *International Tables for X-ray Crystallography* (edited by J.A. Ibers and W.C. Hamilton, Kynoch, Birmingham 1974) pp. 71
16. D.Y. Smith: Phys. Rev., A **35**, 1381 (1966)
17. S. Miyake: *X-ray Diffraction* (Asakura, Tokyo 1969)
18. R.J. Weiss: *X-ray Determination of Electron* (Distributions, North-Holland, Amsterdam 1966)
19. D.T. Cromer: Acta Crystallogr., **18**, 17 (1965)
20. W. Kohn and L.J. Sham: Phys. Rev., A **140**, 1133 (1965)
21. J.A. Bearden: Rev. Mod. Phys., **31**, 78 (1967)
22. H. Brysk and C.D. Zerby: Phys. Rev., **171**, 292 (1968)
23. L. Gerward, G. Theusen, M.S. Jensen, I.Alstrup: Acta Crystallogr., A **35**, 852 (1979)
24. M. Stibius-Jensen: Phys. Lett., A **74**, 41 (1979)
25. S.C. Roy, R.H. Pratt and L. Kissel: Rad. Phys. Chem., **41**, 725 (1993)
26. R.H. Pratt, L. Kissel and P.M. Bergstrom Jr. in: *Resonant Anomalous X-ray Scattering*, ed. by G. Metrlik, C.J. Sparks and K. Fischer (North-Holland, Amsterdam 1994) pp. 9
27. L. Kissel and R.H. Pratt: Acta Crystallogr., A **46**, 170 (1990)
28. M.S. Wang: Phys. Rev., A **34**, 636 (1986)
29. D. Creagh: Nucl. Instrum. Meth. Phys. Res., A **295**, 417 (1990)
30. S. Hosoya: Rigaku-Denki-J., **10**, 7 (1979)
31. B.K. Teo: *EXFAS Basic Principles and Data Analysis* (Springer, Berlin, Heidelberg, New York 1986)
32. P.A. Lee, P.H. Citrin, P. Eisenberger and B.M. Kincaid: Rev. Mod. Phys., **53**, 761 (1981)
33. B. Lengeler in: *Resonant Anomalous X-ray Scattering*, ed. by G. Metrlik, C.J. Sparks and K. Fischer (North-Holland, Amsterdam 1994) pp. 35
34. U. Bonse and M. Hart: Appl. Phys. Lett., 7(1965),238.; Z. Phys., **189**, 151 (1966)
35. U. Bonse and G. Meterlik: Z. Phys., B **24**, 189 (1976)
36. U. Bonse and J. Hellkötter: Z. Phys., **223**, 345 (1969)
37. D.C. Creagh and M. Hart: Phys. Status Solidi, B **37**, 753 (1970)
38. M. Hart: Proc. Roy. Soc. Lond. A **346**, 1 (1975)
39. U. Bonse, I. Hartmann-Lotsch, H. Lotsch, K. Olthoff-Mienter: Z. Phys., B **47**, 297 (1982)
40. J.J. Hoyt, D. de Fontaine and W.W. Warburuton: J. Appl. Crystallogr., **17**, 344 (1984)
41. U. Bonse and I. Hartman-Lotsch: Nucl. Instrum. Met. Phys. Res., **222**, 185 (1984)
42. W.K. Warburton and K.F. Ludwig Jr.: Phys. Rev., B **33**, 8424 (1986)
43. F. Stanglmeir, B. Lengeler, W. Weber, H. Göbel and M. Schuster: Acta Crystallogr., A **48**, 626 (1992)

44. R. Begum, M. Hart, K.R. Lea and D.O. Siddons: Acta Crystallogr., A **42**, 456 (1986)
45. W.H. Zachariasen: Acta Crystallogr., **18**, 714 (1965)
46. S. Hosoya and T. Yamagishi: J. Phys. Soc. Jpn., **21**, 2638 (1966)
47. N.N. Sirota: Acta Crystallogr., A **25**, 223 (1969)
48. T. Kawamura and T. Fukamachi: Jpn. J. Appl. Phys. Suppl., **17-2**, 228 (1978)
49. K. Shinoda, K. Sugiyama and Y. Waseda: High Temp. Mater. Process, **14**, 75 (1995)
50. SCM-AXS(Tohoku): available at http://www.iamp.tohoku.ac.jp/; http://www.tagen.tohoku.ac.jp/
51. P. Fuoss and A. Bienenstock in:*Inner-Shell and X-ray Physics of Atoms and Solids*, ed. by D.J. Fabian, KH. Kleinpoppen and L.M. Watson (Plenum Press, New York 1981) pp. 875
52. R.C. Lye, J.C. Phillips, K. Kaplan, S. Doniach and K.O. Hodgdon: Proc. Nat. Acad. Sci. USA, **77**, 5884 (1980)
53. M. Hart and D.P. Siddons: Proc. Roy. Soc. Lond., A **376**, 465 (1981)
54. C. Cusaitis and M. Hart: Proc. Roy. Soc. Lond., A **354**, 291 (1977)
55. P. Dreier, P. Rabe, W. Malzfeldt and W. Nieman: J. Phys. C. Solid State, **17**, 3123 (1984)
56. A.P. Wilkinson and A.K. Cheetham: J. Appl. Crystallogr., **25**, 654 (1992)
57. P. Suortti, J.M. Hastings and D.E. Cox: Acta Crystallogr., A **41**, 413 (1985)
58. G. Will, N. Masocchi, M. Hart and W. Parrish: Acta Crystallogr., A **43**, 677 (1987)
59. H. Cole and N.R. Stemple: J. Appl. Phys., **33**, 2227 (1962)
60. T. Fukamachi and S. Hosoya: Acta Crystallogr., A **31**, 215 (1975)
61. T. Fukamachi, S. Hosoya, T. Kawamura and M. Okunuki: Acta Crystallogr., A **35**, 104 (1979)
62. T. Saka and N. Kato: Acta Crystallogr., A **43**, 252 (1987)

# 5. In-House Equipment and Synchrotron Radiation Facilities for Anomalous X-ray Scattering

An energy lower than the absorption edge is usually selected in anomalous X-ray scattering (AXS) measurement, in order to avoid strong fluorescent radiation from a sample. In this energy region, the real part of the anomalous dispersion factor changes drastically, whereas the imaginary part and its energy variation are quite small. However, as long as only characteristic radiation produced from the sealed tube X-ray source is used, the change due to the anomalous dispersion effect is limited for a small number of elements (see Fig. 3.2) and is on the order of a few percent of the normal atomic scattering factor. This is because the energies of the characteristic radiation are not always close enough to the absorption edge for a desired element. Therefore, the AXS measurements are designated to vary the X-ray energy continuously by combining an appropriated crystal monochromator with white radiation, such as a tungsten or gold target with a rotating-anode-type X-ray generator, or a synchrotron-radiation source. In this way, by tuning the crystal monochromator, the anomalous dispersion can be maximized.

## 5.1 In-House Equipment

The schematic of the optics for the in-house AXS facility at the Institute of Multidisciplinary Research for Advanced Materials, Tohoku University, is shown in Fig. 5.1 [1, 2]. A photograph is shown in Fig. 5.2. In this equipment, the rotating-anode X-ray generator (Rigaku Rotaflex RU-300), with a tungsten or gold target, is employed as a white-radiation source. The incident X-ray energy can be selected in the region of 8 to 13.5 keV by using a Johansson-type crystal monochromator ($25 \times 50 \times 0.3\,\mathrm{mm}^3$, $R = 320$ mm); for example, Ge (111), Ge(220) or Si(200) crystals, have a bandpass width of about 10 eV. The fluxes of the incident X-ray beam to the sample are obtained over $10^6$ photons/s. Figure 5.3 shows the flux of the incident X-ray beam obtained from this in-house equipment [2]. An energy up to 36 keV can also be obtained by using the higher-order reflections from the Ge(111) or Ge(220) crystal, although the power of the incident X-ray beam is reduced by less than two orders of magnitude. It is worth mentioning that the incident X-ray beam intensity is monitored by measuring the temperature diffuse scattering from a thin aluminum foil placed in the incident beam pass or by

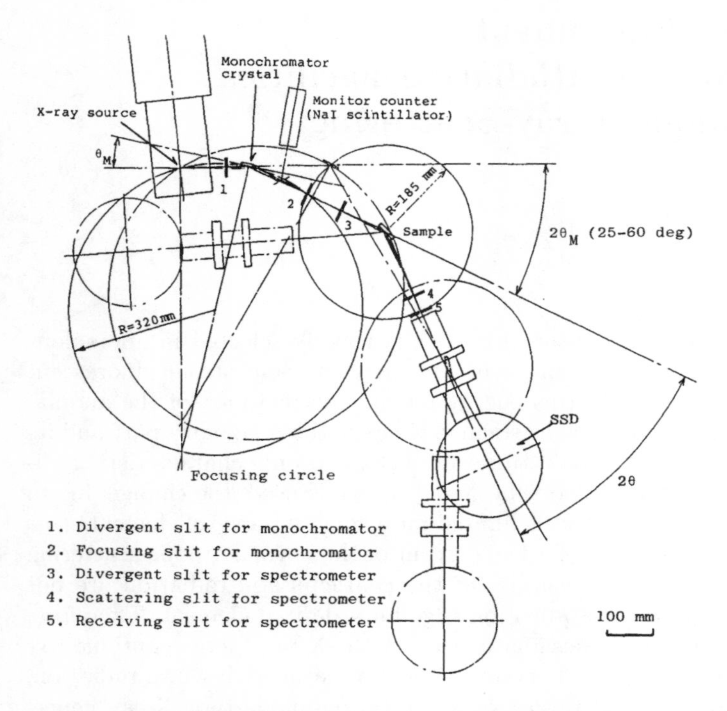


**Fig. 5.1.** Schematic of the optics of the in-house AXS equipment, Tohoku University [1,2]

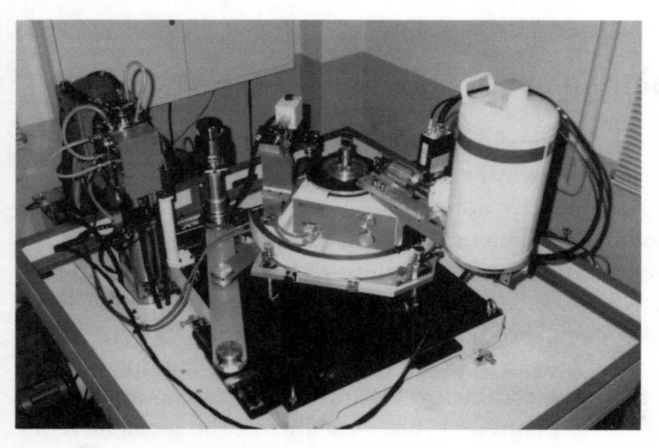

**Fig. 5.2.** Photograph of the in-house AXS equipment, Tohoku University [1]

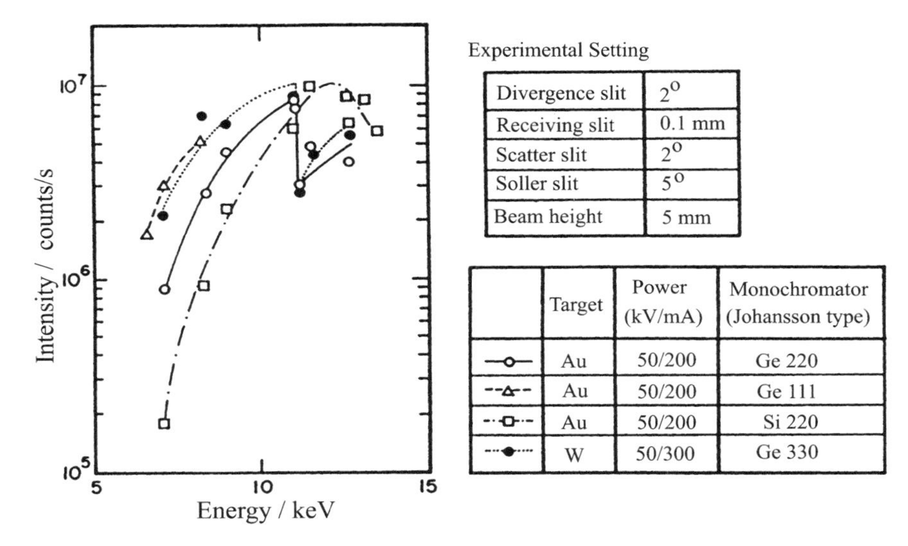


| Divergence slit | $2^{o}$ |
|---|---|
| Receiving slit | 0.1 mm |
| Scatter slit | $2^{o}$ |
| Soller slit | $5^{o}$ |
| Beam height | 5 mm |

| | Target | Power (kV/mA) | Monochromator (Johansson type) |
|---|---|---|---|
| —o— | Au | 50/200 | Ge 220 |
| --△-- | Au | 50/200 | Ge 111 |
| --□-- | Au | 50/200 | Si 220 |
| ···●··· | W | 50/300 | Ge 330 |

**Fig. 5.3.** Measured power of the incident X-ray beam as a function of energy from the in-house AXS equipment, Tohoku University, operated under several conditions [2]

a $N_2$ gas flow type ion chamber placed in front of the sample. The sample chamber and the X-ray beam path are filled with He gas to reduce air scattering and also to avoid the reduction of the incident beam intensity by air absorption.

In the ordinary AXS measurement, an appropriate energy near the absorption edge of interest is turned from a white X-ray source by using a crystal monochromator. Also, we employ an another method, where white X-ray radiation is used directly and the AXS intensities at desired energies are chosen from an energy spectrum recorded at a certain scattering angle by an energy-sensitive, solid-state detector (SSD) [3]. The entire AXS profiles are obtained by moving the detector around the sample. The scanning of the SSD around a sample is a particular feature of this method, different from the energy-dispersive X-ray diffraction (EDXD) method that has already been applied to many structural analyses of liquids and glasses [4,5,6].

Figure 5.4 is a schematic of the experimental set up. This system consists of a rotating-anode X-ray generator of 18 kW with a molybdenum target and a pure germanium SSD mounted on a double-axis diffractometer. When a large amount of incident photons are available in a very short time, the so-called pile-up phenomenon of pulse output from the SSD can induce spurious peaks and background radiation in the higher-energy region. Therefore, needless fluorescent radiation from a sample should be eliminated.

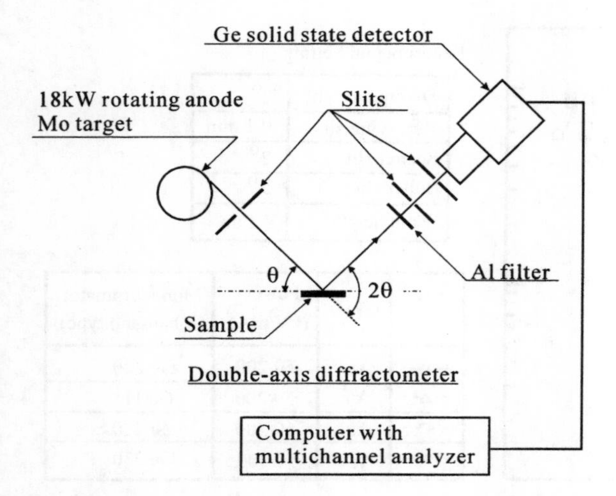


**Fig. 5.4.** Schematic of the experimental set up for the in-house AXS apparatus [3]

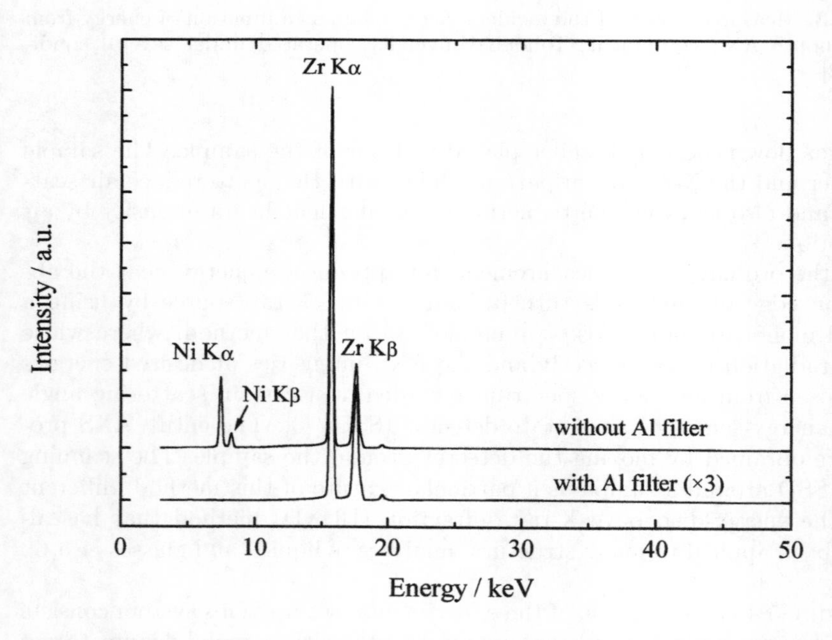


**Fig. 5.5.** Energy spectra diffracted at 40° in $2\theta$ with and without an aluminum plate of 0.8 mm thickness placed in front of the detector [3]

Figure 5.5 shows the energy spectra diffracted at 40° in $2\theta$ with and without an aluminum plate of 0.8 mm thickness placed in front of the detector, using the results of the analysis of amorphous $Zr_{60}Al_{15}Ni_{25}$ alloy as an example. The strong fluorescence of Ni from the sample can be sufficiently reduced in the energy region above 15 keV usually employed in the experiment with a rotating-anode X-ray generator. Then the AXS profiles at various scattering angles are obtained by collecting intensities at appropriate energies with respect to the absorption edge of a specific element, Zr (18.00 keV) in the present case.

The AXS measurements using synchrotron radiation are usually carried out at the incident energies of 50 and 300 eV below the absorption edge of a desired element, where the real part of the anomalous dispersion term drastically changes. On the other hand, we should rather select, using the in-house AXS equipment described in Fig. 5.4, two incident energies where the anomalous dispersion terms change slowly as a function of energy. Two energies of $E_1 = 19.6$ keV and $E_2 = 25.0$ keV were selected for the amorphous $Zr_{60}Al_{15}Ni_{25}$ alloy case [3]. The reasoning is as follows: The energy resolution of about 200 eV of the SSD is much larger than that of a few eV attained by the single-crystal monochromator with synchrotron radiation. Nevertheless, the different contributions from the anomalous dispersion term at the two energies of $E_1 = 19.6$ keV and $E_2 = 25.0$ keV enable us to provide the environmental structure function around Zr by taking the difference between these two AXS intensity profiles.

Figure 5.6 shows the coherent X-ray scattering intensities of the amorphous $Zr_{60}Al_{15}Ni_{25}$ alloy in electron units per atom measured at incident energies of 19.6 and 25.0 keV; they are 1.6 and 7.0 keV above the Zr K absorption edge, respectively. As clearly seen at the bottom of Fig. 5.6, a quite sizable difference in intensity is detected. This can be attributed to the difference in contribution from the anomalous dispersion effect of Zr. There are differences in detail when compared with the case of synchrotron radiation [7], but the over all agreement is surprisingly good.

The following comments should also be made: Considering that the energy resolution of the SSD is about 200 eV, the resultant energy resolution, $\Delta E$, is 600 eV. Therefore, the resolution of the scattering wave vector, $\Delta Q$, can be expressed as

$$Q + \Delta Q = \frac{4\pi(E + \Delta E)}{hc^{\circ}} \sin\left(\frac{\theta}{2}\right), \tag{5.1}$$

where $h$ is Planck's constant, $c^{\circ}$ is the speed of light and $\theta$ is the scattering angle. Equation (5.1) can be rewritten as a function of $Q$:

$$\Delta Q = 0.024Q. \tag{5.2}$$

In the relatively-low-$Q$ region, the change in scattering intensity of X-rays for non-crystalline materials is known to be distinct. This includes the first and

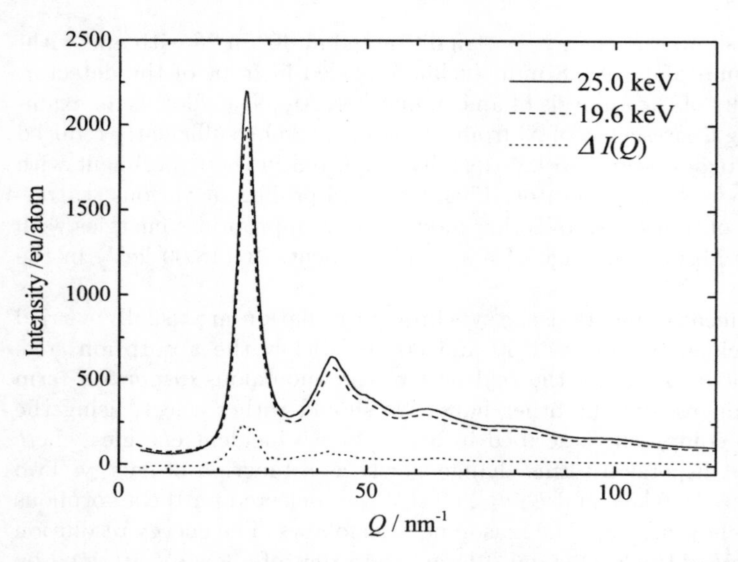


**Fig. 5.6.** Coherent intensity profiles of the amorphous $Zr_{60}Al_{15}Ni_{25}$ alloy in electron units per atom measured at the incident energies of 19.6 and 25.0 keV [3]

second peak region. Thus, the resolution of $Q$ in the low-$Q$ region needs to be sufficiently high to reproduce the variation of the X-ray scattering intensity. From (5.2), $\Delta Q$ around the first peak ($Q = 25$ nm$^{-1}$ in the amorphous $Zr_{60}Al_{15}Ni_{25}$ alloy case) is estimated to be 0.6 nm$^{-1}$. Therefore, the $Q$ resolution in this type of experimental set up is fully adequate for reproducing the intensity variation of the first peak of the amorphous $Zr_{60}Al_{15}Ni_{25}$ alloy. With an increasing scattering wave vector, the $Q$ resolution becomes worse. However, this inconvenience is not so significant because the variation of scattering intensity is drastically reduced with an increasing wave vector, mainly arising from the typical structural feature of an almost complete loss of positional correlation at a few nearest-neighbor distances away from any origin in non-crystalline materials. In addition, the present result covers the scattering intensity profile up to 180 nm$^{-1}$, in comparison to the ordinary case (150 nm$^{-1}$) with Mo K$\alpha$ radiation. This is known to improve the resolution in the radial distribution function (RDF) calculated by the Fourier transform of the interference function.

The most significant problem with the in-house AXS equipment appears to be that the incident beam intensity is still not sufficiently strong for the general purpose of the AXS measurements. With a current intensity of the order of $10^6$ photons/s, it takes about two weeks for the non-crystalline sample case to obtain a good intensity profile, because each measurement of scattering intensity should be repeated several times scanning in order to reduce the statistical error. This condition is not so convenient for the measurement of high-temperature melts, due to experimental difficulties arising from the high chemical reactivity of a liquid sample. In addition, considering the stability of the X-ray generator presently employed, this is the maximum continuous operation time for full power. Therefore, it will be helpful to increase the power of the X-ray generator from 18 to 21 kW and also to mount the position-sensitive proportional counter (or the imaging plate counting system) on the spectrometer [7].

Incidentally, the power of the monochromated X-ray beam obtained from the synchrotron radiation source is of the order of $10^9$ to $10^{10}$ photons/s. In this regard, it is obvious that the availability of intense white X-rays from a synchrotron-radiation source has greatly improved the quality of the AXS measurements.

## 5.2 Synchrotron Radiation Facility

In this section, the synchrotron radiation facility for the AXS measurements is described by using the cases carried out at the beam lines of station A2 in the Cornell High Energy Synchrotron Source (CHESS) Laboratory, Ithaca, New York, and of stations 6B and 7C in the Photon Factory (PF-KEK), Institute of Materials Structure Science, High Energy Accelerator Research Organization, Tsukuba [8].

From the intense white radiation, a monochromatic, incident X-ray beam at energies ranging from 4 to 21 keV, in the case of PF-KEK, can be obtained with an Si(111) double-crystal monochromator. Its optimum energy resolution is about 5 eV at 10 keV [9]. The incident beam is monitored by an $N_2$ gas flow type ion chamber placed in front of the sample so as to keep constant a certain preset number of photons in this ion chamber at each angular position. Even though an energy level that is lower than the absorption edge is selected in the AXS measurements, some fluorescent radiation is frequently emitted from the sample. This arises mainly from the tail of the band path and the higher harmonic diffraction of the crystal monochromator is not negligible. This is particularly true when the incident X-ray energy is tuned close to the absorption edge.

The effect of the higher harmonics can be reduced by intentionally detuning the second crystal in the double-crystal monochromator with a piezoelectric device attached to it, although about one-fourth to one-fifth of the intensity of the first-order diffraction is lost. The ratio of integrated intensities of the higher harmonics to the first-order scattering from the sample appears to be reduced to less than 0.5%. The fluorescent radiation intensity from both of these two origins changes with time as the monochromator crystal is heated up or as the electron path shifts in the storage ring. For this reason, the separation of such a fluorescent component from the scattered intensity is crucial for keeping sufficient reliability of the AXS measurement. In order to overcome this difficulty, the use of an energy-sensitive, solid-state detector

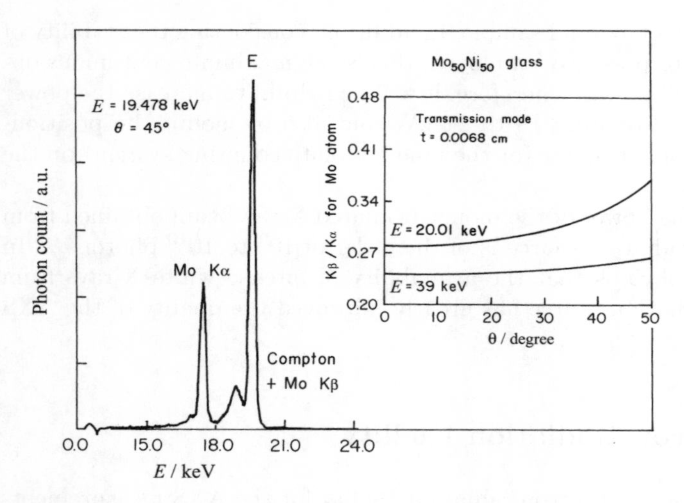


**Fig. 5.7.** Typical energy spectrum of scattered X-ray photons measured for the amorphous $Mo_{50}Ni_{50}$ alloy near the K edge of Mo by the Ge-SSD [12]. The $K\beta/K\alpha$ ratio calculated for the tail of monochromator bandpass ($E = 20.01$ keV) and for the second harmonics ($E = 39$ keV).

(SSD) appears to overcome problems encountered with the usual scintillation counter. Of course, the fluorescent radiation can be discriminated by the diffracted-beam crystal analyzer (see, for example, [10]), but this results in a severe reduction in intensity and thus decreases the statistical accuracy of the results [11].

On the other hand, the typical energy resolution of about 300 eV at 20 keV of a SSD can discriminate most of the fluorescent radiation and also minimize the sensitivity to the high harmonic component of the incident X-rays. Figure 5.7 shows a typical energy spectrum on the multi-channel analyzer obtained by an intrinsic Ge detector for the measurement of an amorphous $Mo_{50}Ni_{50}$ alloy [12]. Even though the energy of incident X-rays was tuned at 19.478 keV, sufficiently below the K absorption edge (20.004 keV) of the Mo atom, the fluorescent lines were quite appreciable. As easily seen from the spectrum in Fig. 5.7, the energy resolution of the SSD is high enough to separate clearly the Mo-$K\alpha$ component from the elastically or inelastically (Compton) scattered photons; whereas, the $K\beta$ component could not be separated by the SSD, although the Compton component can be evaluated by the theoretical values (see, for example, [13]) with the so-called Breit–Dirac recoil factor [14]. However, the contribution from the $K\beta$ component can be numerically subtracted in the data-reduction process, so long as the $K\alpha$ component is monitored by a single channel analyzer (SCA) during the course of the experiment, because the intensity ratio of $K\beta/K\alpha$ is relatively easy to determine through the experiment (so-called off-Bragg condition) [12] or through the theoretical evaluation [15, 16]. It is worth mentioning that this intensity ratio is almost independent of the excitation energy predicted theoretically. However, it depends upon the scattering angle and the diffraction mode, transmission or reflection (see, for example, Fig. 5.7). It should be noted that one of the disadvantages of an SSD is the so-called pulse pile-up problem, when a large amount of photons are available in a very short time. The use of a pulse-pile-up rejecter, with the counting rate below about $10^4$/s, and dead-time correction are necessary for overcoming this problem.

There are two possible diffraction modes for liquid samples by applying the AXS method with synchrotron radiation. One is the transmission geometry, the other is the reflection geometry. Since each mode has its own advantages and disadvantages, a better diffraction geometry should be selected by considering the physicochemical properties of the liquid samples of interest. It may safely be said that the transmission geometry with a devised cell is suitable for a liquid sample which has a relatively small mass-absorption coefficient and high vapor pressure. On the other hand, the reflection geometry from the free surface of a liquid is preferred in cases where the sample has a large mass-absorption coefficient and lower vapor pressure components. Some essential points of this particular facility are described below.

### 5.2.1 Transmission Mode

Figure 5.8 shows a schematic of the experimental set up in the transmission mode using a devised quartz cell. The sample cell is made of quartz glass and the molten sample, for example, CuBr or RbBr [17, 18], can be filled in a cell through tubes with a spacing of 50 μm sandwiched between two windows (10 × 20 mm$^2$ and 100 μm thickness). The optimum sample thick-

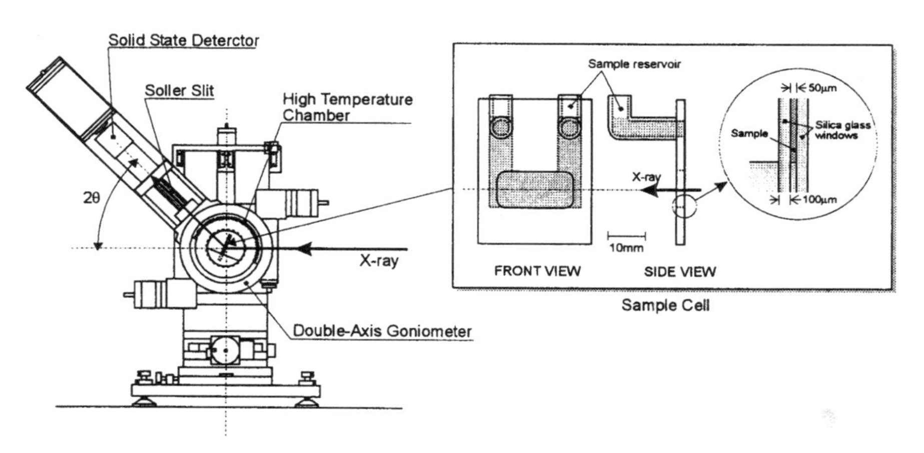


**Fig. 5.8.** Schematic of the experimental set up used for the AXS measurements using a devised quartz cell in the transmission mode [17]

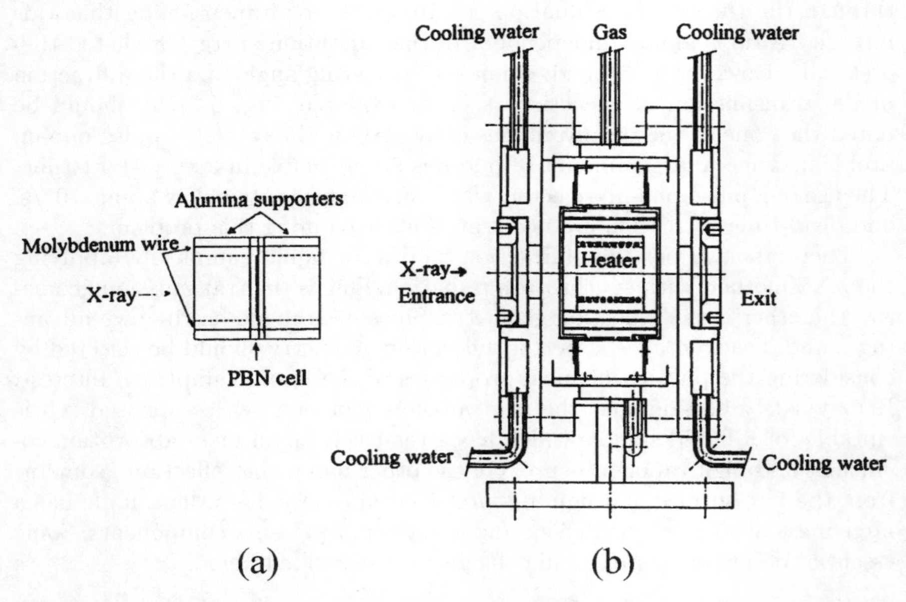


**Fig. 5.9.** Schematic of a devised sample cell of PBN and a furnace assembly for high-temperature melts near 1700 K [19]

ness (50 μm in the present case) was determined from a condition that the scattered intensity has to be sufficiently secured and the absorption is not too strong. The sample is heated to a temperature of about 1000 K by a heating element of Kanthal wire in a purified nitrogen atmosphere. It is suggested that the sample reservoir is encased with nickel for trapping molten salt vapor from the reservoir. This sample cell was positioned in the center of a high-temperature chamber provided with a water-cooled system. For the measurements at temperatures above 1200 K, another devised cell, as illustrated in Fig. 5.9, must be used [19]. In this case, the sample cell is made of high-purity pyrolytic boron nitride (PBN, Shin-Etsu Chemical Co. Ltd.), because of its excellent physical and chemical properties, such as low X-ray absorption, low porosity and its applicability to temperatures up to 2000 K, with good resistance against corrosion. As shown in Fig. 5.9, the sample is sandwiched between two circular plates of PBN and fixed by two cylindrical supports made of alumina, which are tightly wound with molybdenum wire.

As shown in Fig. 5.8, the present facility involves the $\omega$–$2\theta$ goniometer originally designed by Ishikawa and his colleagues [20], as a part of the goniometer system for precision diffractometry and topography. Some modification is required to mount a handy type of SSD and receiving and scattering slit holders with a fine height adjustment on the counter arm. An SSD filled with liquid nitrogen weighing about 6 kg could be used without any prob-

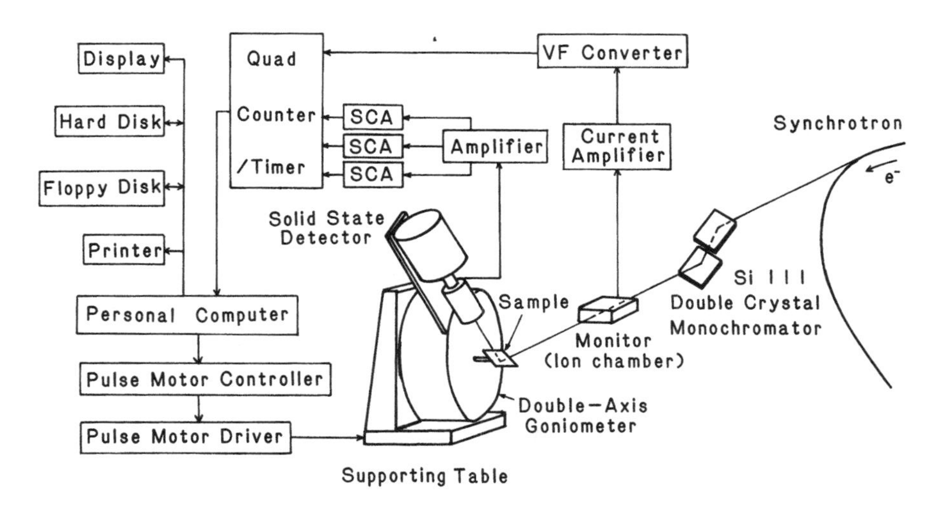


**Fig. 5.10.** Schematic of the experimental set up including the associated electronics system for the AXS measurements using synchrotron radiation

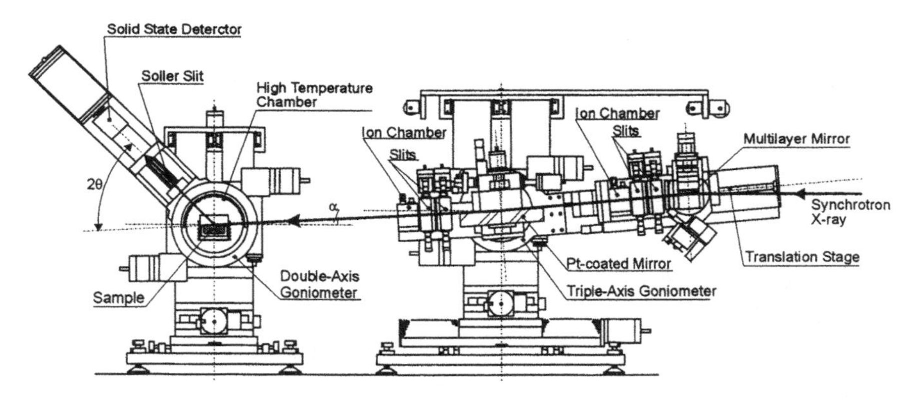


**Fig. 5.11.** Schematic of the AXS measurements for a liquid sample in the asymetrical reflection mode with synchrotron radiation [21]

lem, because this goniometer system is designed to hold more than 10 kg in weight. The $\omega$ and $2\theta$ axes allow 360° rotation and are independently driven by stepping motors with the finest step of rotation of 0.0002° for the $\omega$-axis and 0.0004 ° for the $2\theta$ axis. This goniometer system is also mounted on the $x$–$z$ translation table provided with the finest step (0.002 mm) both for translation and adjusting the sample position in the beam line. In order to set the goniometer surface parallel to the beam, this $x$–$z$ translation stage can be manually rotated around the $z$-axis with the range of $\pm 3°$ in a horizontal plane by adjuster screws. For those interested in conducting AXS measurements, a schematic of the experimental set up including the associated electronics system is shown in Fig. 5.10.

### 5.2.2 Reflection Mode

In order to maintain the free surface of a liquid sample, the diffractometer requires asymmetrical reflection optics, where the incident beam of the synchrotron radiation is irradiated to the horizontal surface at a fixed small glancing angle. A schematic and a photograph of a newly developed glancing-angle control system [21] are shown in Figs. 5.11 and 5.12, respectively. A schematic drawing of the X-ray optics is also given in Fig. 5.13, explaining this particular apparatus [22]. A monochromatic and horizontal X-ray beam extracted from the continuous synchrotron spectrum by a double-crystal Si(111) monochromator is first bent downwards through an angle $2\phi(E)$ by a W/Si multi-layer mirror ($2d = 4.05$ nm). The multi-layer mirror is set on a circular stage ($\phi$ rotation) which can be translated along a 600 mm-long translation stage. Both of these stages are mounted on the arm rotating around the $2\phi$-axis, which is one of the axes of the triple-axis goniometer. The three motions of these stages are synchronously controlled so that the diffracted beam from the multi-layer mirror just passes onto the $2\phi$-axis. Then, the diffracted beam is totally reflected by a flat fused quartz mirror coated with platinum, which is mounted on the center of the triple-axis goniometer and can be rotated around the $\omega$-axis. This mirror can be used for further fine adjustment to keep the direction of the emitted beam constant; a fixed glancing angle, $\alpha$, is used throughout the available X-ray energy region. When $2\phi$ is larger than $\alpha$, the two mirrors are arranged in the $(+,-)$ setting on the triple-axis goniome-

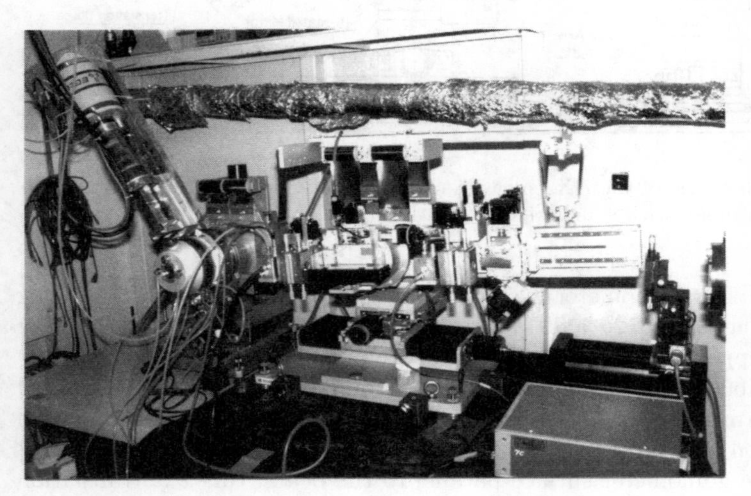

**Fig. 5.12.** Photograph of the AXS measurements for a liquid sample in the asymmetrical reflection mode with synchrotron radiation [21]

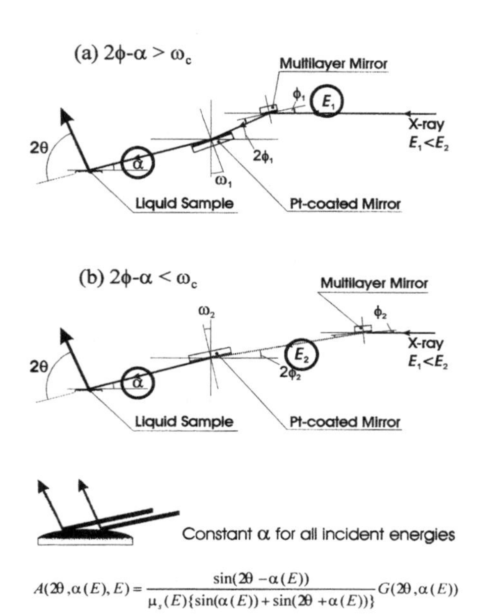


$$A(2\theta,\alpha(E),E)=\frac{\sin(2\theta-\alpha(E))}{\mu_s(E)\{\sin(\alpha(E))+\sin(2\theta+\alpha(E))\}}G(2\theta,\alpha(E))$$

**Fig. 5.13.** Schematic of the X-ray optics for the asymmetrical reflection mode with synchrotron radiation [22]

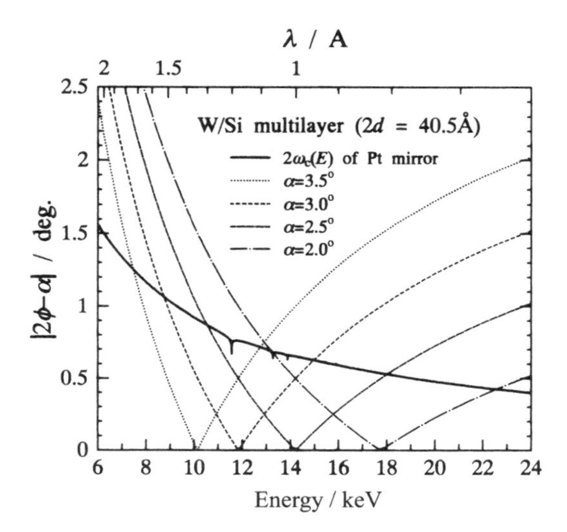


**Fig. 5.14.** Relationship between the angle correction, $|2\phi(E) - \alpha|$, for adjusting the emitted-beam direction and the critical angle, $2\omega_c(E)$, of total reflection with the Pt-coated mirror [21]

ter and otherwise in the (+,+) setting. Figure 5.14 describes the relationship between the angle correction, $|2\phi(E)-\alpha|$, for adjusting the emitted-beam direction and the critical angle, $2\omega_c(E)$, of total reflection with the Pt-coated mirror, at which the reflectivity can be expected to be about 50%. So long as $2\omega_c(E)$ is larger than $|2\phi(E)-\alpha|$, it is possible to adjust the emitted-beam direction without any significant loss of intensity. For example, it was found that the beam direction can be kept constant through an energy range from 8.5 to 15 keV by using the present mirrors when the glancing angle of $\alpha$ is chosen to be 3.0° [21].

The high-temperature chamber is mounted at the center of the double-axis diffractometer placed downstream of the apparatus described previously (see the left-hand side of Fig. 5.11). This chamber has a 15-mm-wide slot with a beryllium window through the water-cooled enclosure wall, in order to allow the passage of X-rays.

## References


1. Y. Waseda: J. Phys. Colloq. C8, Suppl., **46**, C8–293 (1985)
2. E. Matsubara and Y. Waseda: *Proc. Inter. Conf. on Rapidly Solidified Materials* (San Diego, Feb. 3–5 1985), (ASM Int. Materials Park, Ohio 1985) pp. 161
3. S. Sato, M. Imafuku, E. Matsubara, A. Inoue and Y. Waseda: Mater. Trans., **42**, 1977 (2001)
4. J.M. Prober and J.M. Schultz: J. Appl. Crystallogr., **8**, 405 (1975)
5. T. Egami: J. Mater. Sci., **13**, 2587 (1978)
6. V. Petkov, S. Takeda, Y. Waseda and K. Sugiyama: J. Non-Cryst. Solids, **168**, 97 (1994)
7. S. Sato, E. Matsubara, Y. Waseda, T. Zhang, A. Inoue: Mat. Res. Soc. Symp. Proc., **554**, 101 (1999)
8. M. Nomura, A. Koyama and M. Sakurai: *KEK Report 91-1* (National Laboratory High Energy Physics, Tsukuba 1991)
9. M. Nomura: *KEK Report Internal 87-1* (National Laboratory High Energy Physics, Tsukuba 1987)
10. P.H. Fuoss, W.K. Warburton and A. Bienenstock: J.Non-Cryst. Solids, **35/36**, 1233 (1980)
11. T. Matsushita and H. Hashizume: *Handbook on Synchrotron Radiation (Vol.1)*, ed. by E.E. Koch (North-Holland, Amsterdam 1983) pp. 263
12. S. Aur, D. Kofalt, Y. Waseda, T. Egami, H.S. Chen, B.K. Teo and P. Wang: Nucl. Instrum. Meth. **222**, 259 (1984)
13. D.T. Cromer and J.B. Mann: J. Chem. Phys., **47**, 1892 (1967)
14. T. Egami in: *Glassy Metals*, ed. by H.J. Güntherodt and H. Beck (Springer, Berlin, Heidelberg, New York 1981)
15. N. Venkateswara Rao, S. Bhuloka Reddy, G. Satyanarayana and D.L. Sastry: Physica, **138c**, 215 (1986)
16. J.H. Schofield: Atom. Data Nucl. Data Tables, **14**, 121 (1974)
17. M. Saito, C.Y. Park, K. Omote, K. Sugiyama and Y. Waseda: J. Phys. Soc. Jpn., **66**, 633 (1997)
18. M. Saito, S.C. Kang and Y. Waseda: Jpn. J.Appl. Phys. Suppl., **38**, 596 (1999)
19. A.H. Shinohara, K. Omote, S. Kawanishi and Y. Waseda: Jpn. J.Appl. Phys., **35**, 2218 (1996)

20. T. Ishikawa, J. Matusi and T. Kitano: Nucl. Instrum. Meth., A **246**, 613 (1986)
21. M. Saito, C.Y. Park, K. Sugiyama and Y. Waseda: J.Phys. Soc. Jpn., **66**, 3120 (1997)
22. M. Saito and Y. Waseda: J. Synchrotron Rdiat., **7**, 152 (2000)

# 6. Selected Examples of Structural Determination for Crystalline Materials Using the AXS Method

The determination of site occupancy in crystalline materials consisting of more than two elements is essential for discussing physical and chemical properties of materials. However, it is not so easy even when using the Rietveld analysis [1], because convergence is often not obtained after a number of iterations. The discovery of new materials such as quasi-crystals [2], ultrafine particles [3] and oxide superconductors [4] has generated new streams of activity in the materials science and engineering community. It is highly desirable to obtain quantitatively accurate information about their atomic structure, because of the many possibilities arising from the combination of more than two components, usually three or four, plus defects.

The use of the anomalous X-ray scattering (AXS) method provides one way of overcoming the experimental difficulties in these subjects, by making available sufficient atomic sensitivity arising from the "anomalous dispersion effect" near the absorption edge or by providing an appreciable difference in the crystallographic structure factors. This applies even for two elements with atomic numbers that are close in the periodic table such as Fe and Ni. For this reason, the usefulness of this relatively new method for structural characterization of crystalline materials is presented with some selected examples in this chapter.

## 6.1 Spinel Ferrites

The unit cell of ferrite spinels, expressed by the general formula $MFe_2O_4$, where M is a divalent cation, contains eight formula units and is usually referred to the space group $Fd3m(O_h^7)$ with cations occupying the special positions of $8a$ and $16d$. The ideal structure has cubic close packing of oxygen ($32e$), where one-eighth of the tetrahedral and half of the octahedral interstices are known to be occupied. For example, $Zn^{2+}$ cations usually prefer to occupy the tetrahedral sites in zinc ferrite of $ZnFe_2O_4$ as a "normal" spinel. Many ferrites have an "inverse" or "partly inverse" structure, where the divalent $M^{2+}$ cations are distributed both in the octahedral and tetrahedral sites.

Magnetic properties of ferrite spinels are known to be sensitive to the cation distribution [5]. However, the determination of the cation distribution in ferrite spinels by the usual X-ray diffraction method is generally not easy, because the X-ray scattering ability of the component M, such as Zn and Ni, is close to that of host element, Fe. The AXS method using energies near the absorption edge of Fe and M is one way to bring about a significant breakthrough by permitting an appreciable difference in the crystallographic structure factors [6, 7]. This is particularly true when coupled with common Rietvelt analysis [1].

The essential points for data processing are described in Chap. 3. The diffracted intensity of the *hkl* reflection is proportional to the square of the so-called crystallographic structure factor, $F_{hkl}$, for the unit cell, and the quantity $F_{hkl}$ is expressed by using three terms for the $8a$, $16d$ and $32e$ sites:

$$\begin{aligned} F_{hkl} = {} & f_{\mathrm{tet}} \exp[B_{\mathrm{tet}}(\sin\theta/\lambda)^2] \sum_{8a} \exp[2\pi\mathrm{i}(hx + ky + lz)] \\ & + f_{\mathrm{oct}} \exp[B_{\mathrm{oct}}(\sin\theta/\lambda)^2] \sum_{16d} \exp[2\pi\mathrm{i}(hx + ky + lz)] \\ & + f_{\mathrm{oxygen}} \exp[B_{\mathrm{oxygen}}(\sin\theta/\lambda)^2] \sum_{32e} \exp[2\pi\mathrm{i}(hx + ky + lz)], \end{aligned} \tag{6.1}$$

where $(x, y, z)$ are the fractional coordinates. $B_{\mathrm{tet}}$, $B_{\mathrm{oct}}$ and $B_{\mathrm{oxygen}}$ are the isotropic temperature factors for the tetrahedral site, the octahedral site and the site for oxygen, respectively. The summation extends over all atoms in the unit cell.

When the AXS measurements are carried out at two energies, $E_1$ and $E_2$, near the absorption edge of a specific element, the following simple but very useful relation can be obtained [8]:

$$r_{hkl,\mathrm{exp}} = \frac{I_{hkl}(E_1)}{I_{hkl}(E_2)} \propto \frac{|F_{hkl}(E_1)|^2}{|F_{hkl}(E_2)|^2}. \tag{6.2}$$

The detected variation in intensity with energy is attributed to the reflections originating only from a specific element, because the anomalous dispersion effects arising from other elements appear to be insignificant in this energy region. Thus, by comparing the measured $r_{hkl,\mathrm{cal}}$ at two energies close to the absorption edge with the value of $r_{hkl,\mathrm{cal}}$ calculated from the model cation distribution in the usual manner, the site occupancy of a specific element can be quantitatively determined. The ratio of crystallographic structure factors calculated at two energies $E_1$ and $E_2$, as a function of the fraction of cations occupying the tetrahedral site in the spinel structure is given in Fig. 6.1, using the results of $ZnFe_2O_4$ as an example [8, 9]. The following comments are worthy of note: In the spinel structure, the intensity ratio of 222 reflection at two energies provides the structural information only on the cations in the octahedral site, whereas the ratio of 422 reflection can be attributed to that of the tetrahedral site. In addition, the intensity ratio defined by (6.2) is set so as to reproduce the calculated ratio $F_{440}(E_1)/F_{440}(E_2)$, because the intensity ratio of 440 reflection is known to be independent of the cation distribution

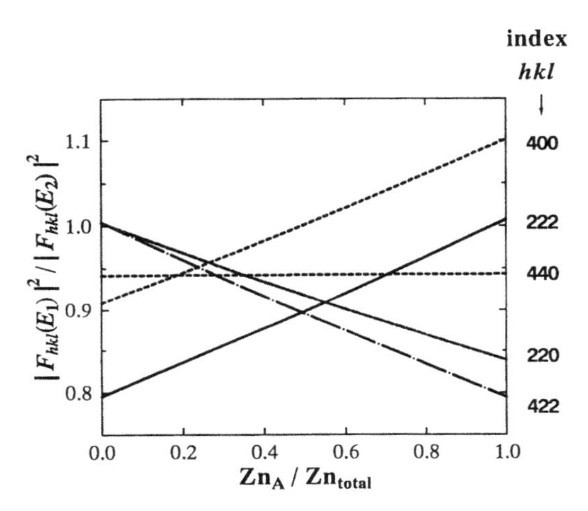


**Fig. 6.1.** The ratio of crystallographic structure factors calculated at two energies, $E_1$ and $E_2$, as a function of the fraction of cations occupying the tetrahedral site in $ZnFe_2O_4$ [8,9]

in the spinel structure. Therefore, the intensity ratios of 222, 422 and 440 reflections are subject to the AXS measurements at energies close to the absorption edges of Fe and M, the element of interest, and this can make the valuable and limited beam time of a synchrotron-radiation source effective.

The $R$-factor given by (3.13) and the iterative procedure (see Fig. 3.7) are introduced in data processing in order to obtain a significant solution [10]. The cation distribution of spinel ferrite samples can be determined so as to minimize the $R$-factor of AXS measurements, $R_{\mathrm{AXS}}$, at energies close to the absorption edges of M and Fe.

It may be helpful to recall the concept of *normal* or *inverse* for site occupancy in the spinel structure. Let us introduce the degree of inversion, $x$, defined by the fraction of $M^{2+}$ cations in the octahedral sites; the scattering factors $f_{\mathrm{tet}}$ and $f_{\mathrm{oct}}$ are described as a function of $x$, as follows:

$$f_{\mathrm{tet}} = (1-x) f_{\mathrm{M}} + x f_{\mathrm{Fe}}, \tag{6.3}$$

$$f_{\mathrm{oct}} = \frac{x}{2} f_{\mathrm{M}} + \frac{2-x}{2} f_{\mathrm{Fe}}. \tag{6.4}$$

It should be noted that $x = 0.0$ and $x = 1.0$ correspond to the particular site occupancies of all cations occupying the tetrahedral site (*normal*) and the octahedral site (*inverse*), respectively.

The results for $ZnFe_2O_4$ and $CoFe_2O_4$ are as follows [10]: The ordinary diffraction patterns for common Rietveld analysis were obtained by using Cu-K$\alpha$ or Co-K$\alpha$ radiation with a pyrolitic graphite monochromator in a diffracted beam. X-ray diffraction patterns of two samples were checked to agree well with the compiled JCPDS data of No. 22-1021 ($ZnFe_2O_4$) and No.

22-1086 ($CoFe_2O_4$). The crystal structures were refined by Rietveld analysis with the generalized computer package program PFLS [11] within the constraints of the chemical formula and full site occupancy. It may be stressed that the lattice constant was determined by the whole powder pattern decomposition method with an internal standard of NBS Si(640*b*) [12]. The values of lattice constant $a$ and the structural parameters $u$ (commonly known as the positional parameter of oxygen) and $B$ determined by Rietveld analysis are summarized in Table 6.1, together with the final values of $R_{\mathrm{Bragg}}$ and $R_{\mathrm{F}}$ [10]. The resultant profile of Rietvelt analysis for the $ZnFe_2O_4$ sample is shown in Fig. 6.2 as an example.

The incident energies used in the AXS measurements are a pair of 6.961 and 7.086 keV, corresponding to 150 and 25 eV below the Fe-K absorption edge (7.111 keV). Similarly, we used two pairs of incident energies 9.511 keV/9.636 keV below the Zn-K absorption edge (9.660 keV) and 7.560 keV/7.684keV below the Co-K absorption edge (7.710 keV). These energies were tuned by using a Si(111) double-crystal monochromator, and its optimum energy resolution was about 5 eV at 10 keV. For convenience, the numerical values of the anomalous dispersion factors used in this work are summarized in Table 6.2.

**Table 6.1.** The crystal structure refinement of $ZnFe_2O_4$ and $CoFe_2O_4$ spinels, using the Rietveld and AXS methods

| | $a$ (nm) | $u$ | $B_{\mathrm{tet}}$ ($\mathrm{nm}^2$) | $B_{\mathrm{oct}}$ ($\mathrm{nm}^2$) | $B_{\mathrm{oxygen}}$ ($\mathrm{nm}^2$) | $R_{\mathrm{Bragg}}$ (%) | $R_{\mathrm{F}}$ (%) |
|---|---|---|---|---|---|---|---|
| $ZnFe_2O_4$ | 0.8442 | 0.2612 | 0.0071 | 0.0044 | 0.0078 | 2.96 | 2.28 |
| $CoFe_2O_4$ | 0.8393 | 0.256 | 0.0056 | 0.0071 | 0.0119 | 3.24 | 2.67 |

**Table 6.2.** Anomalous dispersion terms $f'$ and $f''$ used in this work

| | Energy (keV) | $f'_{\mathrm{Fe}}$ | $f''_{\mathrm{Fe}}$ | $f'_{\mathrm{Co}}$ | $f''_{\mathrm{Co}}$ | $f'_{\mathrm{Zn}}$ | $f''_{\mathrm{Zn}}$ | $f'_{\mathrm{O}}$ | $f''_{\mathrm{O}}$ |
|---|---|---|---|---|---|---|---|---|---|
| Fe-AXS | 6.961 | −3.50 | 0.49 | −2.10 | 0.57 | 0.21 | 2.44 | 0.06 | 0.04 |
| | 7.086 | −5.09 | 0.47 | −2.26 | 0.55 | 0.17 | 2.39 | 0.06 | 0.04 |
| Co-AXS | 7.560 | −2.07 | 3.53 | −3.59 | 0.49 | | | 0.05 | 0.04 |
| | 7.684 | −1.78 | 3.45 | −5.24 | 0.48 | | | 0.05 | 0.04 |
| Zn-AXS | 9.511 | −0.21 | 2.44 | | | 3.80 | 0.50 | 0.03 | 0.02 |
| | 9.636 | −0.17 | 2.38 | | | 5.68 | 0.49 | 0.03 | 0.02 |
| Co-K$\alpha$ | 8.048 | −3.37 | 0.49 | −2.07 | 0.57 | −1.14 | 0.89 | 0.06 | 0.04 |

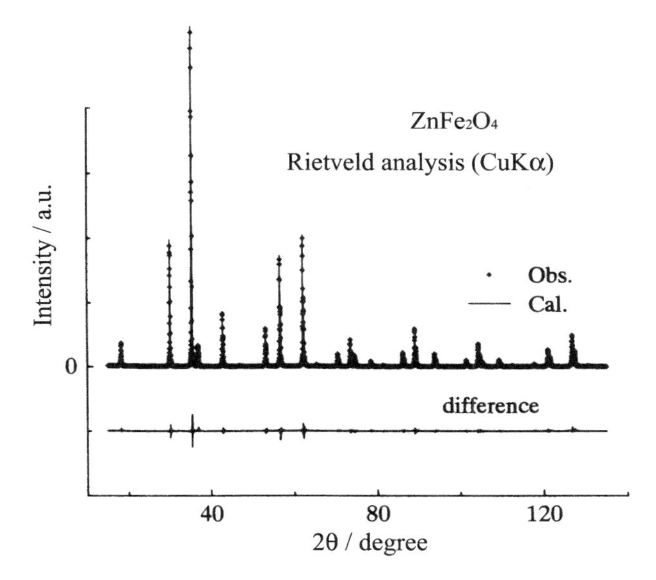


**Fig. 6.2.** X-ray diffraction patterns of $ZnFe_2O_4$. *Top*: the calculated profile is compared with measured values indicated by *crosses*. *Bottom*: the plot of the difference between the two [8]

As easily seen in the results of Fig. 6.3, a significant intensity variation is clearly detected in the 222 reflection at the Fe-K absorption edge, but it is not in the 422 reflection. On the other hand, the reverse behavior is clearly detected in the AXS measurements at the Zn-K absorption edge. This strongly supports the fact that the $ZnFe_2O_4$ sample is classified into the normal type, where $Zn^{2+}$ cations are quite likely to be distributed in the tetrahedral site

**Table 6.3.** The values of $r_{hkl,\mathrm{exp}}$ obtained by the present AXS measurements, $r_{hkl,\mathrm{cal}}$, calculated from the model cation distribution, and the resultant inversion parameter, $x$, for $ZnFe_2O_4$ and $CoFe_2O_4$

| | | $ZnFe_2O_4$ | | $CoFe_2O_4$ | |
|---|---|---|---|---|---|
| | Reflection | $r_{hkl,\mathrm{exp}}$ | $r_{hkl,\mathrm{cal}}$ | $r_{hkl,\mathrm{exp}}$ | $r_{hkl,\mathrm{cal}}$ |
| Fe-AXS | 422 | 1.02 | 1.00 | 0.83 | 0.83 |
| | 222 | 0.40 | 0.41 | 0.66 | 0.67 |
| Co-AXS | 422 | | | 0.98 | 0.98 |
| | 222 | | | 0.88 | 0.86 |
| Zn-AXS | 422 | 0.76 | 0.79 | | |
| | 222 | 1.02 | 1.01 | | |
| Inversion parameter, $x$ | | 0.00 | | 0.78 | |

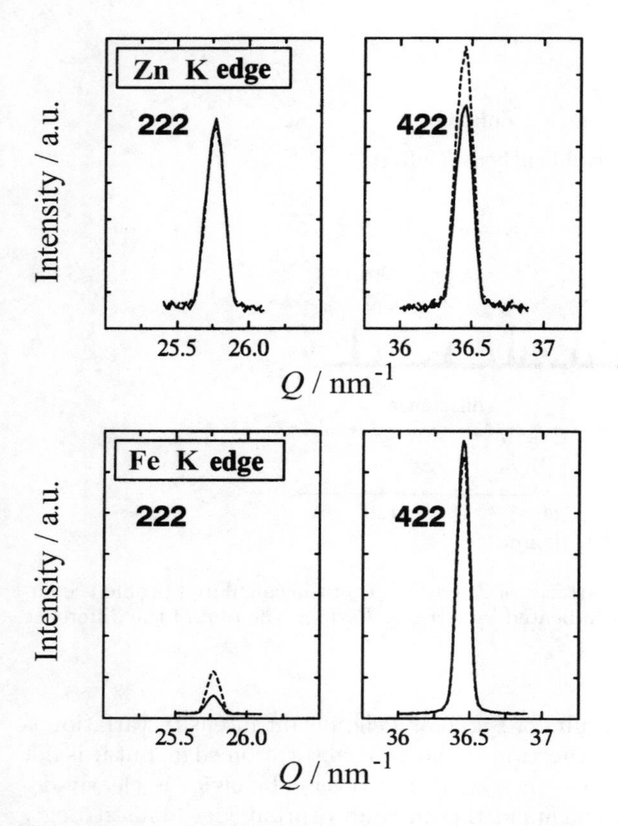


**Fig. 6.3.** The diffraction peak intensities of $ZnFe_2O_4$ measured with two energies at the absorption edges of Fe and Zn. Fe-K edge: *solid* (7.086 keV); *dashed* (6.961 keV). Zn-K edge: *solid* (9.636 keV); *dashed* (9.511 keV) [10]

and $Fe^{3+}$ cations are octahedrally coordinated. The anomalous dispersion effects of Fe and Zn on the intensity ratio at two energies can be calculated as a function of the degree of inversion $x$ (see Fig. 6.1). The results for $x$, which give the minimum value of $R_{\mathrm{AXS}}$ defined by (3.13), are listed in Table 6.3 together with the values of $r_{\mathrm{hkl,exp}}$ and $r_{\mathrm{hkl,cal}}$. These facts quantitatively confirm the concept that all $Zn^{2+}$ cations occupy the tetrahedral site in the $ZnFe_2O_4$ sample presently investigated.

Similar structural analysis was made for the $CoFe_2O_4$ sample. Figure 6.4 shows the energy variation of two reflection peak intensities of $CoFe_2O_4$ obtained from the AXS measurements. The energy dependence is, more or less, detected in all cases measured at the Fe-K and Co-K edges. This implies a disordered distribution of Co and Fe cations between the tetrahedral and octahedral sites. The measured intensity ratio can be reproduced only when using $x = 0.78$. Thus, it may be safely concluded that almost one-forth of the

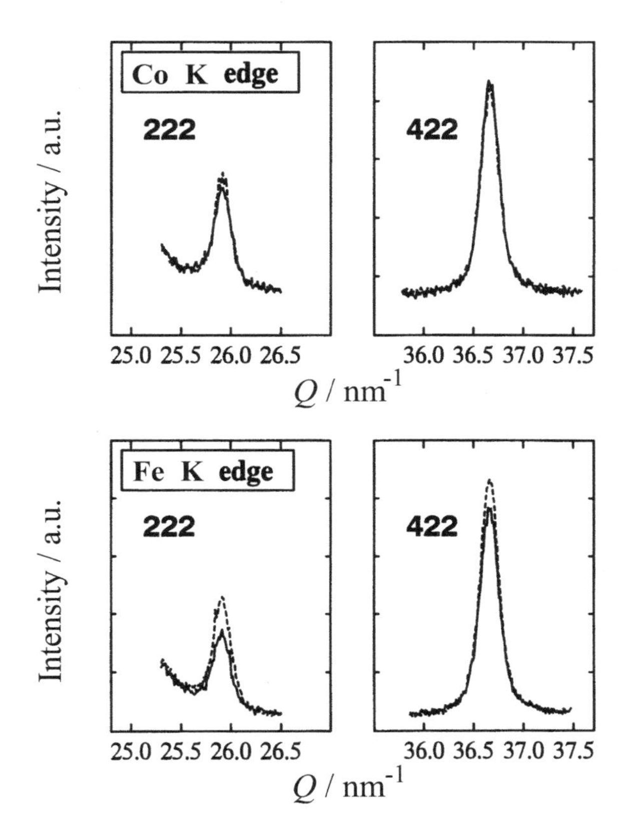


**Fig. 6.4.** The diffraction peak intensities of $CoFe_2O_4$ measured with two energies at the absorption edges of Fe and Co. Fe-K edge: *solid* (7.086 keV); *dashed* (6.961 keV). Co-K edge: *solid* (7.684 keV); *dashed* (7.560 keV) [10]

tetrahedral site are occupied by $Co^{2+}$ cations in the present $CoFe_2O_4$ sample. It should also be stressed that this is the first quantitative information on the $Co^{2+}$ cation distribution in $CoFe_2O_4$ found using X-ray diffraction technique, to the best knowledge of the author.

According to the discussion of O'Neill and Navrotsky [13], the lattice constant $a = 0.838$ nm and the oxygen parameter $u = 0.255$ can be estimated when the inversion parameter is equal to 0.78. These values agree surprisingly well with the results of Tables 6.1 and 6.3. However, it should be kept in mind that the lattice constant of ferrite samples is known to depend frequently upon sample preparation, and the variation of cation distribution with heat treatment is also suggested from the magnetic property measurements of $CoFe_2O_4$ [14]. Such points can be answered by making available systematic structural data from AXS measurements on samples prepared by various conditions as a function of temperature or oxygen fugacity.

In conclusion, the usefulness of the improved structural analysis using the AXS method coupled with Rietveld analysis has been confirmed. Neighboring elements in the periodic table cannot be obviously distinguished with the conventional X-ray diffraction method. However, the AXS method enables us to provide a possible way for solving such a difficulty, as has been demonstrated with the results for $CoFe_2O_4$. Similar results have also been reported for $ZnFe_2O_4$, $NiFe_2O_4$ and $NiAl_2O_4$ samples using an in-house AXS facility [15]. It would be very interesting to extend the AXS method to other spinel-type materials in order to obtain a clearer delineation of the direct link between site occupancy and their characteristic properties.

## 6.2 Superconductors

The discovery of superconductivity above 30 K in the Ba–La–Cu–O system first reported by Bednorz and Müller in 1986 [4] is well known to have generated an enormous amount of relevant activities. Many subsequent investigations indicated several cuprite oxides including rare elements, such as Y, Ba, Sr and Bi, have a superconducting transition temperature higher than the liquid nitrogen temperature (77 K) [16], and it would be very desirable to extend these new materials to practical use [17]. The growing technological and scientific importance of oxide superconductors has led to an increasing need for further studies including structural characterization, in order to describe their atomic-scale structure and to clarify their particular properties at high temperature. However, it is not an easy task to determine the structural parameters of multi-component systems quantitatively accurately, because of the many possibilities arising from the combination of more than two components, usually four components, plus defects in oxide superconductors. The AXS measurement can also provide the local environmental structure around a specific element in a multi-component system, and the results of a $YBa_2Cu_3O_{7-x}$ superconductor by the AXS method are given below.

The preliminary X-ray diffraction results suggest the sample to be an orthorhombic phase, as shown in Fig. 6.5, and the estimated lattice parameters are $a = 0.3886\,\mathrm{nm}$, $b = 0.3820\,\mathrm{nm}$, and $c = 1.1675\,\mathrm{nm}$ using a least-squares fitting method [18]. On the other hand, Fig. 6.6 shows the energy variation in intensity obtained from the AXS measurements at energies of 8.829 and 8.955 keV, corresponding to $-150$ and $-25$ eV away from the Cu-K absorption edge (8.979 keV). Here, the energy variation of $\Delta I_{\mathrm{Cu}}(Q)$ is taken from the difference between $I(-150\text{ eV})$ and $I(-25\text{ eV})$, and the detected energy variation should be attributed to the anomalous dispersion effect of copper.

The oxygen-deficient perovskite atomic arrangement in the orthorhombic crystal structure illustrated in Fig. 6.7 proposed by Izumi et al. [19] is considered as one of the most realistic model structures for the $YBa_2Cu_3O_{7-x}$

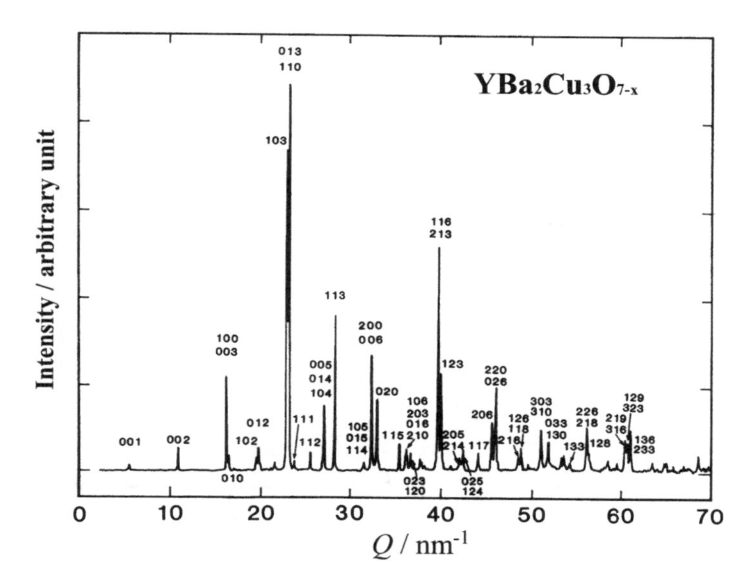


**Fig. 6.5.** Intensity pattern as a function of wave vector of a $YBa_2Cu_3O_{7-x}$ superconductor [18]

superconductor. Thus, the anomalous dispersion effect of copper on the crystallographic structure factor, $F$, can be readily computed from this model structure in the usual way. The signs (+) and (−) in Fig. 6.6 denote such calculation. The numerical examples are as follows:

$$\begin{aligned} \Delta F_{\mathrm{Cu}} &= |F|_{-150\,\mathrm{eV}} - |F|_{-25\,\mathrm{eV}} \\ &= 56.78 - 62.25 = -5.47 && \text{for 100 reflection;} \\ \Delta F_{\mathrm{Cu}} &= 51.93 - 57.12 = -5.19 && \text{for 003 reflection;} \\ \Delta F_{\mathrm{Cu}} &= 158.50 - 153.22 = +5.28 && \text{for 100 reflection;} \\ \Delta F_{\mathrm{Cu}} &= 167.69 - 162.41 = +5.28 && \text{for 013 reflection;} \\ \Delta F_{\mathrm{Cu}} &= 165.53 - 159.99 = +5.54 && \text{for 110 reflection.} \end{aligned}$$

The measured intensity is proportional to the crystallographic structure factor. However, it is also known to depend upon other quantities, such as multiplicity and temperature factors. Considering such points, the agreement in sign between experiment and calculation in Fig. 6.6 for the energy variation arising from the anomalous dispersion effect of copper is rather surprisingly good. On the other hand, some trial calculations with respect to the sign of the energy variation in intensity by replacing Y or Ba for Cu sites in the model structure of Fig. 6.7 provide only gross disagreement between the measured intensity data and calculated values, as shown in Table 6.4. In addition, some other models, such as the tetragonal atomic arrangements with $a = 0.386$ nm and $c = 0.1975$ nm for $(Y, Ba)_3Cu_2O_x$ [20], have been found not to reproduce the characteristic energy dependence of the intensity. For example, the value

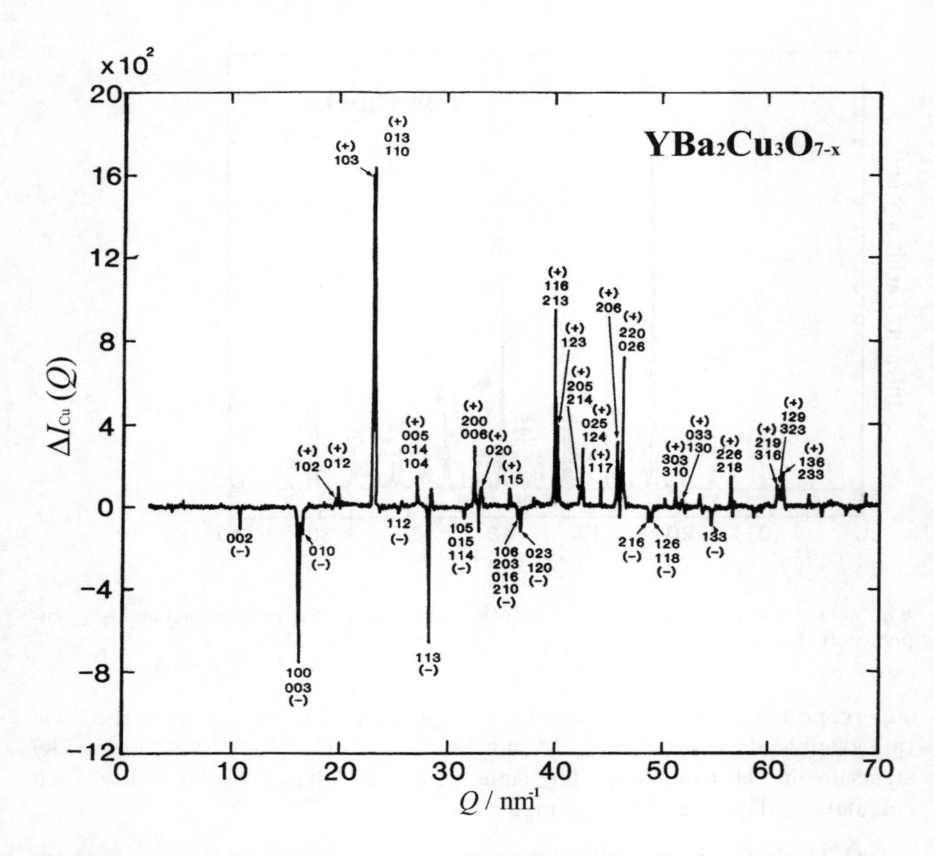


**Fig. 6.6.** Energy variation in intensity of $\Delta I_{Cu}$ for a $YBa_2Cu_3O_{7-x}$ superconductor measured at energies of 8.829 and 8.955 keV, corresponding to −150 and −25 eV away from the Cu-K absorption edge (8.979 keV) [18]

of $\Delta|F_{Cu}|^2$ for 006 reflection ($d = 0.323$ nm), which would be observed near 102 reflection ($d = 0.323$ nm) or 012 reflection ($d = 0.320$ nm), is estimated to be $(93.17) - (100.04)^2 = -1353.4$. Similarly, the value of $\Delta|F_{Cu}|^2 = (79.92)^2 - (78.95)^2 = +154.1$ is obtained for 008 reflection ($d = 0.247$ nm) near 112 reflection ($d = 0.247$ nm). However, the experimental results show the positive sign for 006 and the negative sign for 008, respectively.

The model structure proposed by Izumi et al. [19] has been confirmed after a number of trials using the Rietveld analysis [21]. In other words, the final conclusion for the model structure, the Izumi model in the present case, could be relatively easy to obtain when coupled with the AXS measurements. This is because the feasibility check of a model structure of interest can be readily and simply made by comparing the coincidence of the sign between

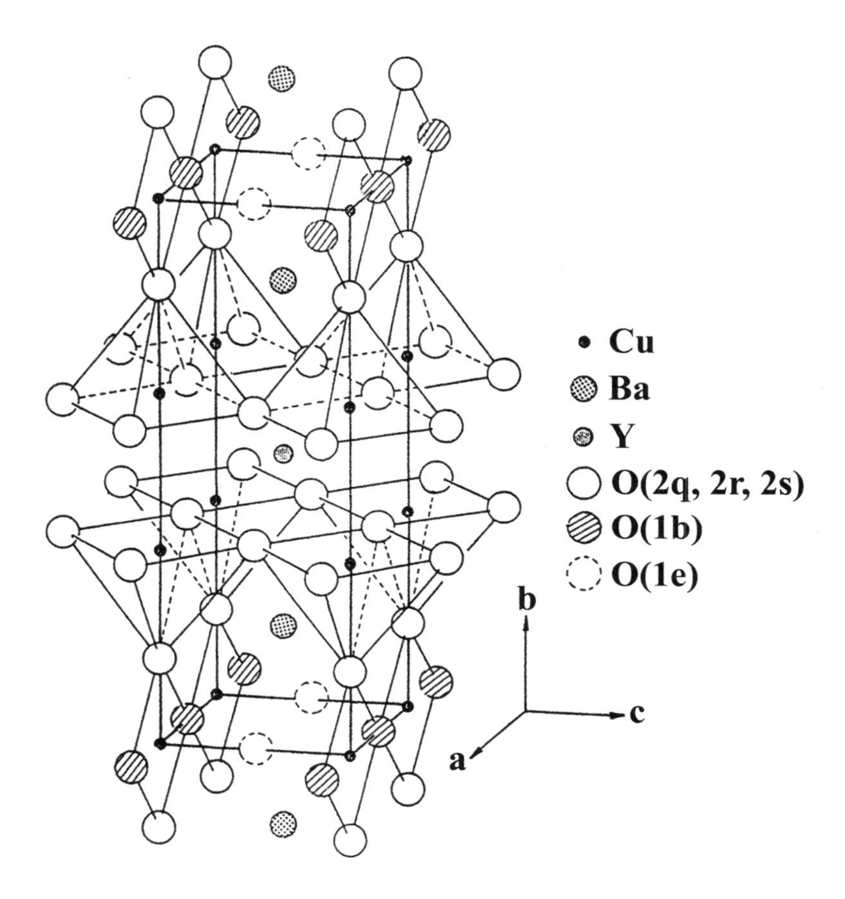


**Fig. 6.7.** Schematic for a model structure of a $YBa_2Cu_3O_{7-x}$ superconductor proposed by Izumi et al. [19]

the AXS data and the calculated values of the crystallographic structure factors for a certain element, as exemplified by the results shown in Fig. 6.6.

Some reservations have been frequently suggested in the past regarding the conclusion of the space group, even for simple binary systems. For example, the crystal structure of a high-temperature phase of copper selenide ($\alpha$–$Cu_2Se$) is classified into the cubic system of the space group denoted by $T_{\rm d}^2$–$F\bar{4}3m$, where four copper ions occupy the $4c$ sites, and the remaining four copper ions are statistically distributed over a lot of equivalent sites of this space group [22]. However, the space group of $O_{\rm h}^5$–$Fm3m$ with a disordered arrangement of all copper ions appears to equally fit the experimental data. For this subject, the AXS measurements provide quite useful information by comparing the coincidence of the sign between the AXS data and calcu-

**Table 6.4.** Comparison of the sign for the energy dependence of the intensity, $\Delta I_{\mathrm{Cu}}(Q)$, between the measured value and the calculation estimated from different atomic positions in the model proposed by Izumi et al. [19]. The signs (+) and (−) describe the sign of the measured and calculated values of $\Delta|F|^2(=|F|^2_{-150\,\mathrm{eV}}-|F|^2_{-25\,\mathrm{eV}})$. *: Disagreement

| $h$ | $k$ | $l$ | Measured | Original Izumi's model | Model by replacing Ba for Cu site | Model by replacing Y for Cu site |
|---|---|---|---|---|---|---|
| 1 | 0 | 0 | (−) | (−) | (−) | (−) |
| 0 | 0 | 3 | (−) | (−) | (−) | (−) |
| 1 | 0 | 2 | (+) | (+) | (−)* | (−)* |
| 0 | 1 | 2 | (+) | (+) | (−)* | (−)* |
| 1 | 0 | 3 | (+) | (+) | (+) | (+) |
| 1 | 1 | 0 | (+) | (+) | (+) | (+) |
| 0 | 1 | 3 | (+) | (+) | (+) | (+) |
| 1 | 1 | 2 | (−) | (−) | (+)* | (−) |
| 1 | 1 | 5 | (+) | (+) | (+) | (−)* |
| 1 | 1 | 7 | (+) | (+) | (−)* | (+) |

lated values of the crystallographic structure factors for a specific element, as exemplified by the results of oxide superconductors.

The results of $\alpha$–$Cu_2Se$ are summarized in Table 6.5 using the AXS measurements at both Cu and Se edges [23]. Here, the energy dependence of $\Delta I_{\mathrm{A}}$(A = Cu or Se) is taken from the intensity at –300 eV away from the absorption edge of the A atom minus the intensity at –25 eV away from the absorption edge of the A atom. The signs of the energy variation in intensity for several reflections (400, 333(+511), 440, 531 and 620 at the Cu edge and 400, 331, 333(+511), 440, 531 and 620 at the Se edge) were found to be positive. On the other hand, the signs of the energy dependence in intensity for 111, 222 and 420 reflections at the Cu edge and 200 reflection at the Se edge were negative. Such particular energy dependence should be attributed to the anomalous dispersion effect of copper or selenium in $\alpha$–$Cu_2Se$. It is apparent from the results given in Table 6.5 that the signs of the energy dependence in intensity obtained by the AXS method agree rather well with the model calculation when the space group of $O_{\mathrm{h}}^3$–$Fm3m$ is employed in the $T_{\mathrm{d}}^2$–$F\bar{4}3m$ case.

The following point is also worthy of note: The environmental radial distribution function (RDF) around copper was calculated from measured energy dependence in intensity of Fig. 6.6 by applying the simple Fourier transforma-

**Table 6.5.** Comparison of the sign of the energy dependence in intensity obtained by the AXS measurements with the calculated values using the different space groups. ×: Not clearly observed

| *hkl* | Cu edge | | | Se edge | | |
|---|---|---|---|---|---|---|
| | Measured | $O_h^5$–$Fm3m$ | $T_d^2$–$Fm\bar{4}3m$ | Measured | $O_h^5$–$Fm3m$ | $T_d^2$–$Fm\bar{4}3m$ |
| 111 | – | −0.57 | 5.16 | + | 16.56 | 19.55 |
| 200 | × | −2.19 | 4.11 | – | −3.16 | −6.09 |
| 220 | + | 32.28 | 56.50 | + | 18.37 | 27.18 |
| 311 | + | 3.90 | 21.72 | + | 10.11 | 17.39 |
| 222 | – | −1.46 | 0.34 | × | 0.30 | −1.20 |
| 400 | + | 2.30 | 6.12 | + | 2.04 | 4.12 |
| 331 | × | −0.39 | 3.40 | + | 0.61 | 4.12 |
| 420 | – | −1.09 | −3.71 | × | 1.71 | −0.39 |
| 422 | + | 2.82 | 24.75 | + | 3.10 | 1.88 |
| 333+511 | + | 1.43 | 16.85 | + | 2.46 | 7.35 |
| 440 | + | 0.52 | 11.82 | + | 0.69 | 4.79 |
| 531 | + | 2.12 | 12.43 | + | 2.37 | 7.87 |

tion, and the resultant RDF is shown in Fig. 6.8 [18]. The mixed contribution from two kinds of Cu-O pairs, one is about 0.190 nm and the other is about 0.230 nm, to the near-neighbor region is suggested by Izumi et al. [19]. As shown in Fig. 6.8, the peak resolution of the present AXS analysis is not sufficiently enough to separate such two Cu-O correlation pairs. Nevertheless, the second peak and its small hump (at about $r$ = 0.42 nm) are in good coincident with the atomic correlation expected in the crystal structure of $YBa_2Cu_3O_{7-x}$. In addition, the orthorhombic structure suggests the region where no atomic correlation at about $r$ = 0.23–0.30 nm, as denoted by the hatched region in Fig. 6.8. The almost zero value observed at about $r$ = 0.27 nm in the environmental RDF for Cu is consistent with such atomic arrangements around copper in the $YBa_2Cu_3O_{7-x}$ superconductor.

## 6.3 Ultrafine Particles and Catalytic Particles

The scattering intensity pattern of ultrafine particles, whose diameters are characterized by their nanometer size, is recognized to be extremely broadened because of the small particle size. For example, the scattering intensity pattern for 4-nm zinc ferrite particles is shown in Fig. 6.9 together with that of the bulk sample for comparison [24]. The positions and relative peak in-

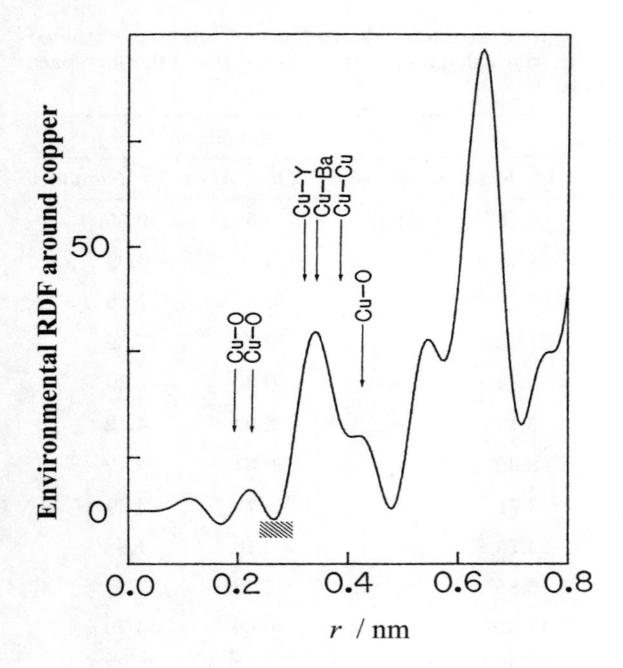


**Fig. 6.8.** Environmental radial distribution function around copper in a $YBa_2Cu_3O_{7-x}$ superconductor [18]. The *arrows* indicate the average distances of atomic pairs estimated from a model structure of Izumi et al. [19]

tensities show no difference between the two samples, except for peak broadening. This suggests that the fundamental atomic structures are identical in both fine particle and bulk samples. It may also be mentioned that the average crystalline size of fine particles can be estimated from the relatively well-separated peaks using the method proposed by Hall [25].

On the other hand, ultrafine particles frequently show some peculiar features distinct from those in the bulk, because of the large fraction of surface atoms. For example, the nanometer-sized zinc ferrite particles show an extremely large magnetization compared to the value reported for the bulk [26]. In order to explain such a characteristic property, the atomic structure of their surface is strongly required. The AXS method is one way to obtain information on the surface structure of ultrafine particles.

The scattering intensity for perfect crystalline particles with a certain size may be calculated using the following simple equation based on Warren's approach [27]:

$$I(Q) = \sum_{j=1}^{N} \sum_{k=1}^{N} f_j f_k \exp(-2M_{\mathrm{DW}}) \frac{\sin(Qr_{jk})}{Qr_{jk}}, \tag{6.5}$$

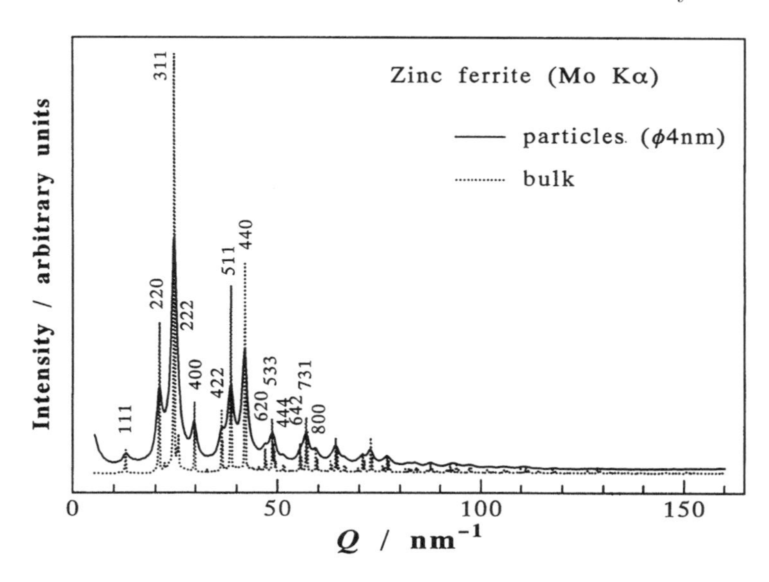


**Fig. 6.9.** Comparison of scattering intensities of 4-nm zinc ferrite particles with those of the bulk sample [24]

where $N$ is the total number of atoms in a crystalline particle, $f_j$ the X-ray atomic scattering factor of element $j$ and $M_{\mathrm{DW}}$ the Debye–Waller temperature factor. The Debye–Waller factor of zinc ferrite particles was estimated from the Debye temperature (630 K), which was estimated from those of ZnO (600 K) and $Fe_2O_3$ (660 K). It should be mentioned here that the contribution of the inter particle interface was excluded in this calculation, because it has been shown to make a significant contribution only in the small-angle region [28].

A scattering intensity calculated from inner atoms of the 4-nm zinc ferrite particles is compared with the experimental intensity data in Fig. 6.10 [24]. The disagreement is clearly detectable. It may be added that the calculated peak intensities come closer to the experimental data when a smaller average particle size is chosen. However, we cannot reproduce the relatively higher background intensity by simply changing the particle size. Thus, another model is required for explaining the experimental data of nanometer-sized zinc ferrite particles. This contrasts with the results of fine gold particles (average particle size = 4 nm), as shown in Fig. 6.11, where the calculated intensity fairly well coincides with the measured one [29]. The fine gold particles consist of a number of perfect single crystals, and no difference is likely to be found in the structure of the surface layer and inner atoms.

An attempt has been made to interpret the difference between two scattering intensities for the 4-nm zinc ferrite particles as observed in Fig. 6.10. The scattering intensity due to atoms inside the particle can be computed

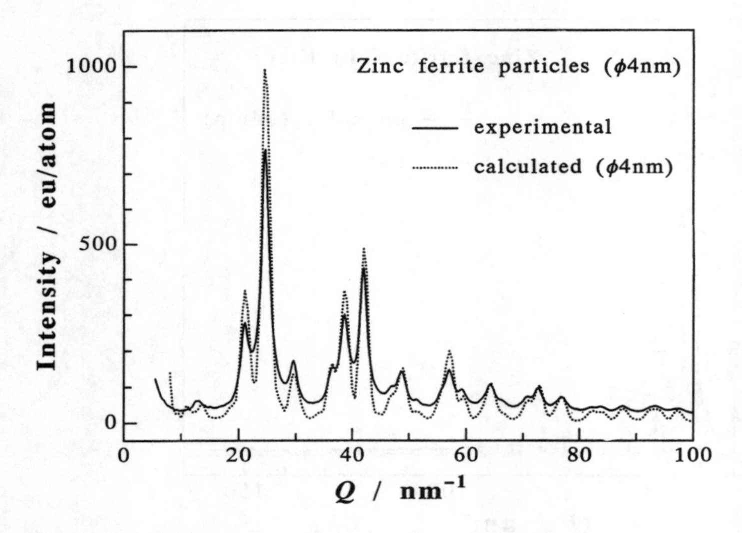


**Fig. 6.10.** Comparison of the measured intensity pattern of 4-nm zinc ferrite particles with the one calculated without considering surface atoms [24]

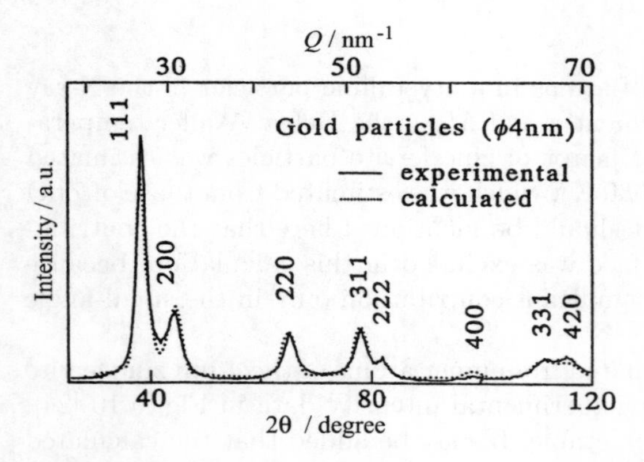


**Fig. 6.11.** Comparison of the measured intensity pattern of 4-nm gold particles with the one calculated without considering surface atoms [29]

as is described with a dotted line in Fig. 6.12, and the difference between the two intensity patterns of experiment and calculation is readily obtained, as provided at the top of figure. This difference profile, corresponding to the contribution from surface atoms of the particles, suggests an amorphous-like intensity pattern consisting of a broad first peak at about 25 $\mathrm{nm}^{-1}$ and second and third broader and weaker peaks at about 40 and 65 $\mathrm{nm}^{-1}$. Assuming that atoms at the surface are randomly displaced with regard to atoms inside the particles, the average phase shift between atoms at the surface

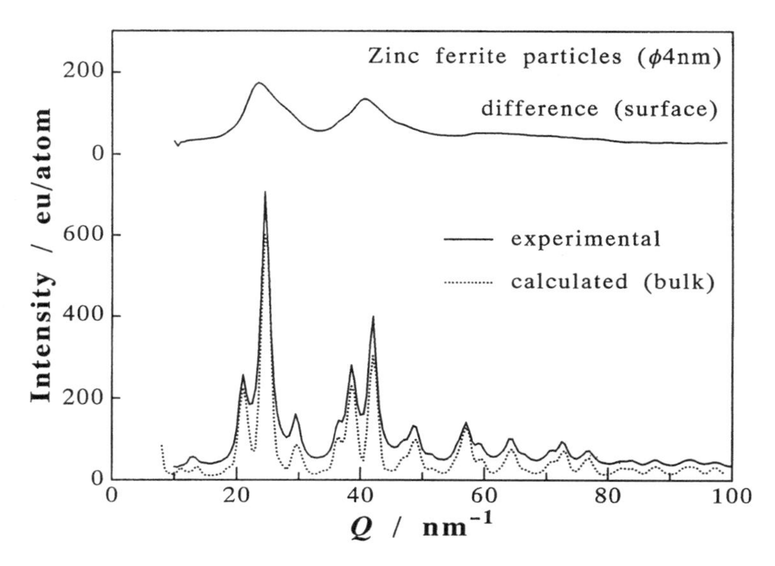


**Fig. 6.12.** The difference (*top*) between the experimental (*bottom*, *solid*) and calculated (*bottom*, *dotted*) intensity profiles for 4-nm zinc ferrite particles [24]

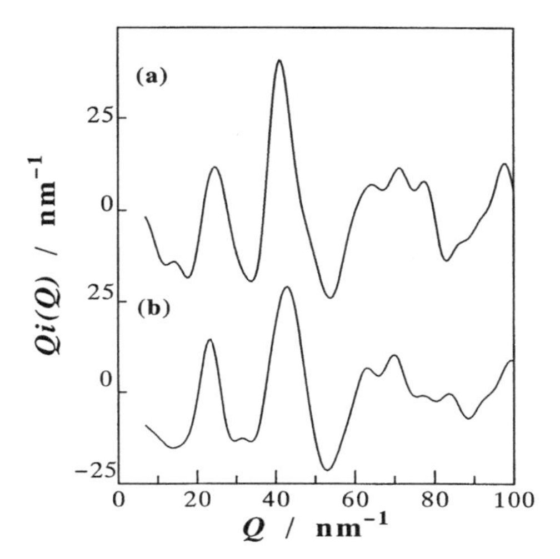


**Fig. 6.13.** Interference functions of the difference between the experimental and calculated intensities **a** for 4-nm zinc ferrite particles and **b** amorphous zinc ferrite film [24, 30]

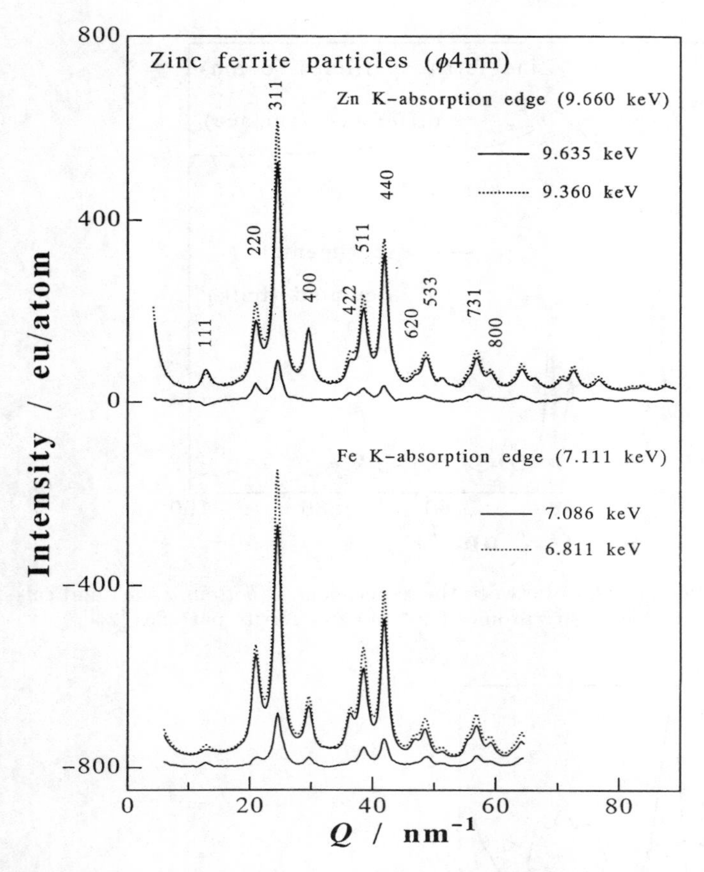


**Fig. 6.14.** Scattering intensity profiles of 4-nm zinc ferrite particles obtained from AXS measurements at the Zn and Fe K absorption edges [24]. The intensity differences are also given beneath the two sets of AXS data

and inside the particle in every direction around the particle becomes zero. Thus, the scattering intensity due to cross term is considered to contribute much less to the total scattering intensity than those for atoms at the surface and inside the particle. Therefore, the difference profile in Fig. 6.12 is attributed to the amorphous-like atomic configuration at the surface. Similarity to the amorphous zinc ferrite film [24, 30] is readily confirmed with the results shown in Fig. 6.13 represented by the interference function form, although there are differences in detail. This suggests that the surface atoms form an amorphous-like structure instead of the zinc ferrite crystalline structure inside the particle. Therefore, let us assume a mono layer surface, with a thickness of about 0.2 nm and a structure different from the bulk.

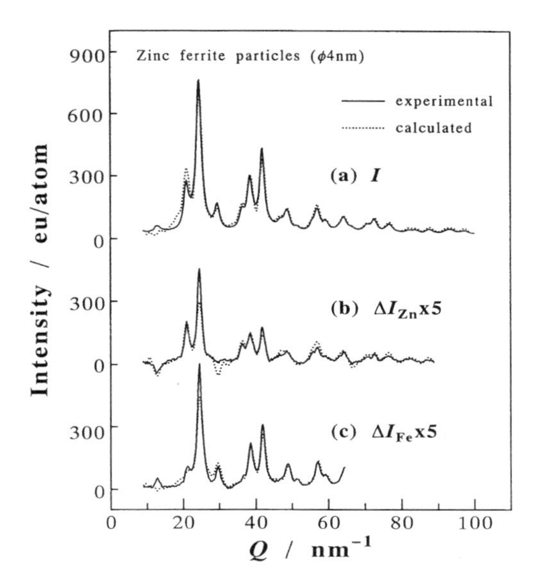


**Fig. 6.15.** The experimental (*solid*) and calculated (*dotted*) intensity profiles obtained **a** by ordinary X-ray diffraction with Mo-K$\alpha$ radiation and **b**,**c** by the AXS measurements below the K absorption edges of Zn and Fe, respectively [24]

In order to confirm such an idea, the AXS measurements for the 4-nm zinc ferrite particles were carried out at the K absorption edges of Fe and Zn [24], and the results are provided in Fig. 6.14. These results are obtained from the measurements at energies of 9.360 and 9.635 keV near the Zn K absorption edge (9.659 keV) and 6.811 and 7.086 keV near the Fe K absorption edge (7.111 keV). The details of the structure of the surface atoms in zinc ferrite particles cannot be revealed from the present results alone; the total scattering intensity in Fig. 6.9 and the AXS intensities obtained at the two edges in Fig. 6.14 were calculated by substituting the scattering intensity of the amorphous zinc ferrite film for the intensity due to the surface component. The calculated intensities are compared with the experimental data in Fig. 6.15. Although the boundary between the surface and bulk components is sharp in the present model, a gradual change from the amorphous-like structure to the crystal might be more plausible. Nevertheless, it may not be overemphasized that good agreement is found by including the background intensities as well as the peak intensities in all three cases. For this reason, the presence of the amorphous-like structure on the surface is considered to be, at least, in the sense of the necessary condition, the most proficient at explaining the scattering intensities for the nanometer-sized zinc ferrite particles by reproducing the three independent scattering intensity profiles described in Fig. 6.15.

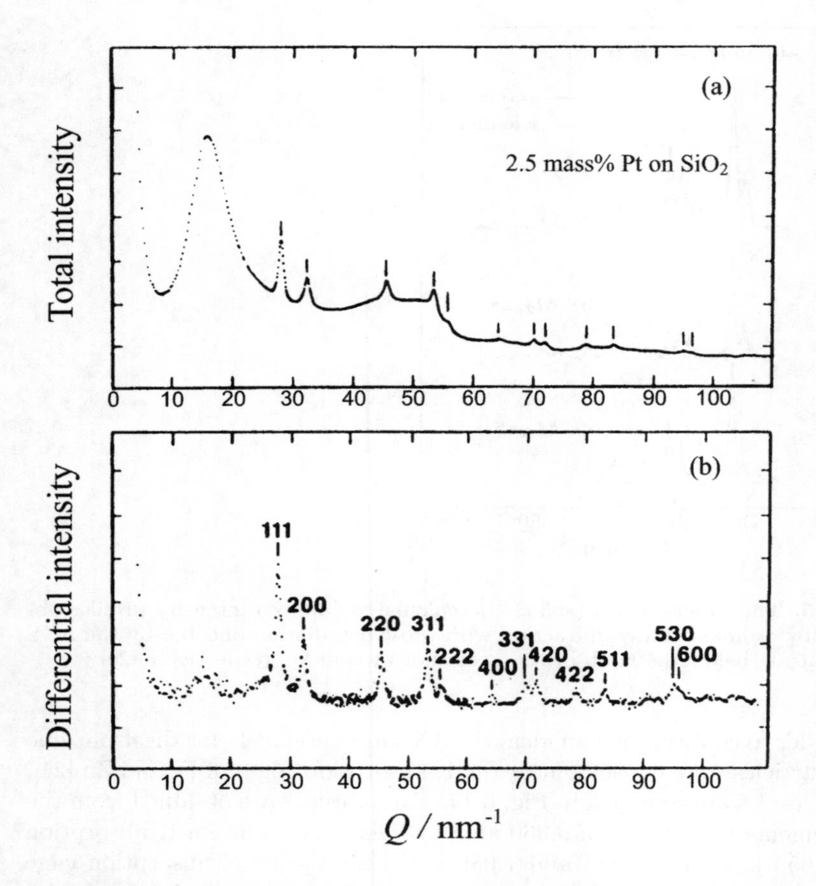


**Fig. 6.16.** **a** Ordinary and **b** differential scattering profiles of 2.5 mass-% platinum particles on silica support obtained by Liang, Laderman and Sinfelt [31]. The differential profile is taken from the two measurements at energies of 99 and 9 eV below the Pt-$L_{III}$ absorption edge

Fine metallic particles on the supported material, such as small Pt particles on silica and small Ni particles on alumina, are successfully used as catalysts. Since the measured intensity data involve contributions from both the support and the catalysts, the intensity from the support material should be accurately subtracted for quantitative structural analysis. As the metallic particle size becomes smaller and smaller, the catalyst peaks broaden, and then the diffuse intensity profile associated with the fine particles prevents us from separating the intensity data of interest. The use of the AXS method provides a breakthrough in this difficulty in a way similar to the cases for a solution sample contained in a cell or a thin film grown on a substrate (see Fig. 3.9). For convenience in further structural studies of the supported catalysts, an example is given below.

Liang, Laderman and Sinfelt [31] reported the AXS measurements of small platinum particles on crystalline alumina, as well as amorphous silica, supports under the name of differential AXS [32]. Figure 6.16**a** shows the scattering profile of 2.5 mass-% platinum particles on silica-glass support obtained from measurement at 99 eV below the white line of the Pt absorption edge, corresponding to the $L_{III}$ absorption edge. The diffraction peaks from platinum particles are found to superimpose with the typical diffuse scattering profile of the amorphous silica support. A similar AXS experiment was made at an energy of 9 eV below the platinum absorption edge. Then, the interesting results of Fig. 6.16**b** were obtained by taking the difference between the two intensity profiles without any serious correction procedure for the support material. The resultant intensity profile clearly indicates the relatively high background, as well as the peak broadening. From the widths of these peaks, the average platinum particle size is estimated to be about 7.5 nm. Although some reservations have been suggested regarding a wide peak detected at about 15 $nm^{-1}$ and the diffuse background in the region of 111 and 200 reflections [31], it would be very promising to extend this AXS approach to structural characterization of a variety of supported catalysts.

## 6.4 Quasi-crystals

The experimental evidence from icosahedral point-group symmetry and long-range orientational order, in rapidly quenched Al–Mn alloys, first reported by Shechtman et al. [2] has stimulated many further theoretical and experimental studies, because of their unusual quasi-periodicity–implying a five-fold symmetry, which is excluded in the conventional crystallographic method. A two-dimensional quasi-crystal with decagonal symmetry has also been reported in some Al-base alloys, such as Al–Fe–Ni and Al–Cu–Co [33, 34], in which the atomic structure has a certain periodicity along the ten-fold axis and a quasi-periodicity in the plane perpendicular to it. Formation of an icosahedral crystal may be related to the structural similarity to an equilibrium phase containing icosahedral clusters in the unit cell. With respect to this subject, Steinhardt et al. [35] found in molecular dynamic simulations that icosahedral correlation increases in the under-cooled liquid state. Such icosahedral ordering is likely to be preserved by rapid quenching. It was also deduced that an icosahedral crystal could condense out of an under-cooled liquid with an atomic structure similar to those in an amorphous alloy [36].

Unlike ordinary crystals, quasi-crystals have an infinite number of sites which are not exactly equivalent. This makes construction of a model structure for quasi-crystals extremely complicated. One of the successful approaches for describing its atomic structure is the "rigid geometry plus decoration" method proposed by Henley [37]. Namely, the atomic structure of a quasi-crystal can be described by placing atoms on a rigid geometrical frame with a certain decoration rule. The use of the anomalous dispersion effect for a specific element is one way to obtain information about the decoration rule in quasi-crystals.

It should be mentioned that the first AXS experiment was reported by Kofalt et al. [38] for the quasi-crystalline $Pd_{58.8}U_{20.2}Si_{20.6}$ alloy using the $L_{III}$ absorption edge of U. The results obtained with atomic pair distribution function (PDF) analysis are summarized as follows: The environmental structure around U in the icosahedral phase of the quasi-crystalline $Pd_{58.8}U_{20.2}Si_{20.6}$ alloy compares favorably with that of a quasi-crystalline lattice decorated with U atoms at the vertices of the primitive quasi-crystalline lattice up to as much as 4 nm [39]. These results were confirmed by Fucks et al. [40]. The AXS method has since been applied to many quasi-crystals, such as $Al_{75}Cu_{15}V_{10}$ [41], $Al_{75}Fe_{15}Ni_{10}$ [42], $Al_{65}Cu_{20}TM_{15}$ (TM = Fe, Ru and Os) [43, 44, 45] and $Al_{65}Cu_{15}Co_{20}$ [46], and some selected results are given below.

A stable single icosahedral phase was prepared by crystallization of an amorphous $Al_{75}Cu_{15}V_{10}$ alloy prepared by rapid quenching from the melt. Figure 6.17 provides the intensity profiles of four samples of the $Al_{75}Cu_{15}V_{10}$ alloy [41]. The fundamental feature of the as-quenched sample given at the bottom of this figure is the typical non-crystalline structure. However, some distinct features can be detected. For example, a pre-peak is found at $Q = 15$ $nm^{-1}$. A marked shoulder at the low-$Q$ side of the main peak and a small hump at $Q = 40$ $nm^{-1}$ are also visible. These similar structural features are frequently observed in some Al-base amorphous alloys, such as Al–Si–Mn [47] and Al–Cu–V–Si–Mo [48]. The presence of a pre-peak has been qualitatively interpreted as one of the compound-forming behaviors, different from the typical random distribution of the constituents. Additional details about the pre-peak are briefly explained in the following.

The schematic of the atomic configuration in the unit cell of crystalline $Al_{45}V_7$ is illustrated in Fig. 6.18 using the results of Brown [49]. The unit cell of crystalline $Al_{45}V_7$ consists of two kinds of icosahedra, formed by Al atoms at vertices and V atoms at their centers, and a polyhedron consisting of two icosahedra sharing one of the pentagonal planes with each other. The average distance between V atoms located at the centers of these icosahedra is about 0.51 nm. The correlation length for a structure as evidence of certain ordering in a non-periodic system may be computed from a simple empirical relation between the correlation length, $r$ in real space, and the peak position, $Q_p$ in the intensity profile: $Q_p \cdot r = 2.5\pi$, as is known in many liquid metals and alloys [50]. The correlation length causing the pre-peak at 15 $nm^{-1}$ was estimated to be 0.52 nm. This value shows remarkable agreement with the distance between V atoms located at the centers of icosahedral clusters in the crystalline $Al_{45}V_7$, although the present samples are of a ternary alloy containing Cu as well as Al and V.

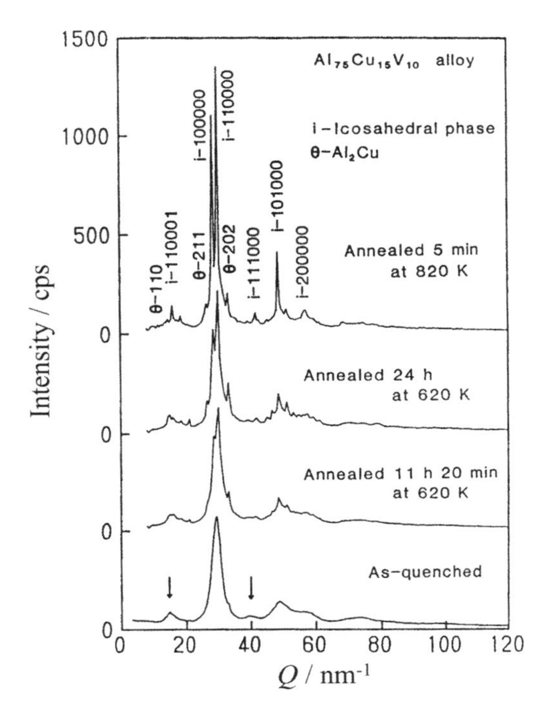


**Fig. 6.17.** Intensity profiles of four samples of the $Al_{75}Cu_{15}V_{10}$ alloy [41]

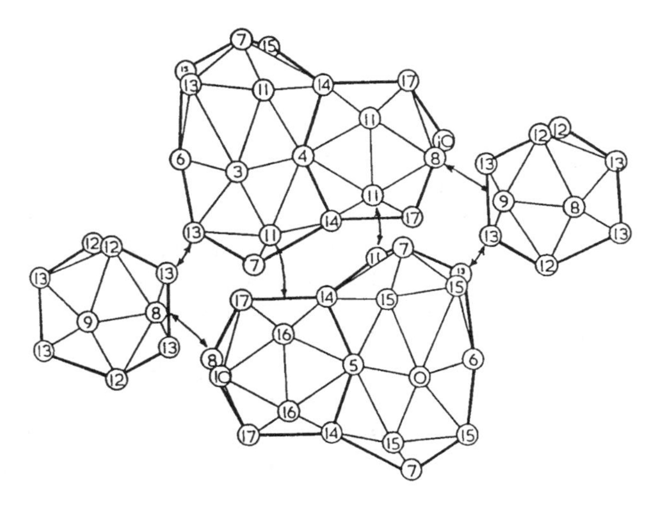

**Fig. 6.18.** Schematic for the atomic configuration in the unit cell of crystalline $Al_{45}V_7$ [49]

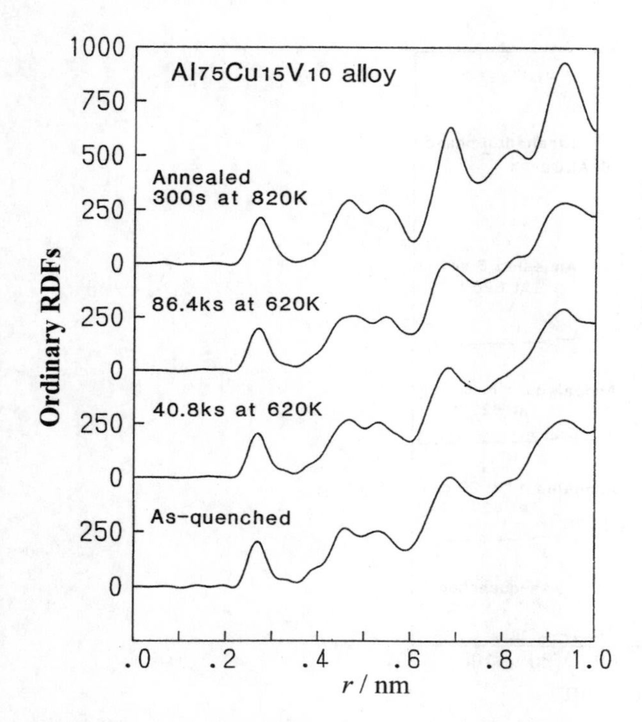


**Fig. 6.19.** Radial distribution functions of four samples of the $Al_{75}Cu_{15}V_{10}$ alloy [41]

On annealing the as-quenched amorphous alloy sample, the main peak and its shoulder separate into two icosahedral peaks, which are 100000 and 110000 according to the indexing scheme proposed by Bancel et al. [51]. Similarly, the small hump observed at $Q = 40$ nm$^{-1}$, the second peak and its shoulder sharpen and grow into the icosahedral peaks 110001, 111000 and 200000, respectively. This continuous growth from diffused peaks in the amorphous state into sharp peaks in the quasi-crystalline state is significant evidence for the structural similarity between the amorphous and icosahedral phases. This is consistent with the RDF results of Fig. 6.19. Nevertheless, the AXS measurements were carried out at the Cu-K absorption edge, in order to obtain information about the location of Cu in these icosahedral clusters. The AXS profiles of the amorphous and icosahedral phases are provided in Figs. 6.20 and 6.21, respectively [41]. In these measurements, the incident X-ray energies are tuned at 8.680 and 8.955 keV, corresponding to 300 and 25 eV below the Cu-K absorption edge (8.979 keV).

As shown in Fig. 6.20, the essential profile of the intensity difference in the amorphous state shows a very similar feature to the original intensity profile before taking the difference. Thus, the Cu atoms are likely to

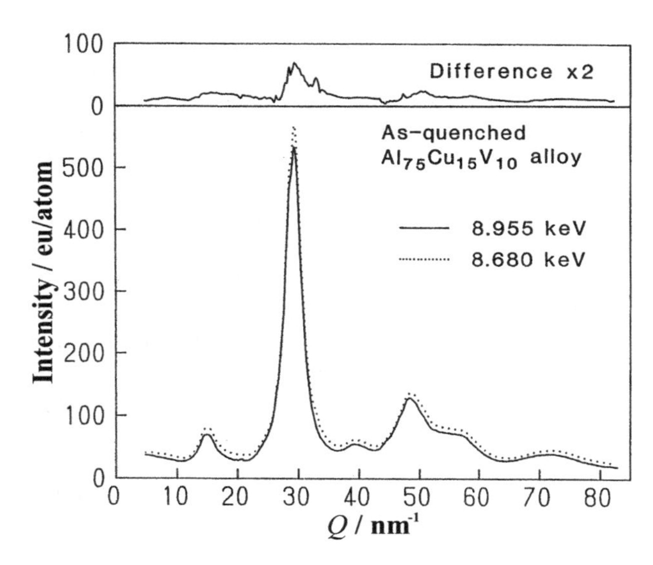


**Fig. 6.20.** Intensity profiles of the amorphous $Al_{75}Cu_{15}V_{10}$ alloy obtained from the AXS measurements at the Cu-K absorption edge [41]. *Top*: intensity difference

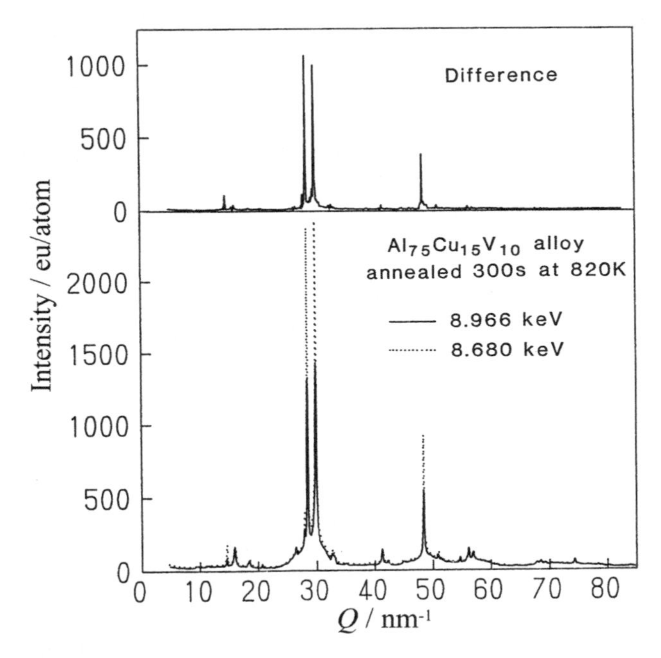


**Fig. 6.21.** Intensity profiles of the icosahedral $Al_{75}Cu_{15}V_{10}$ alloy obtained from the AXS measurements at the Cu-K absorption edge [41]. *Top*: intensity difference

be homogeneously distributed in the icosahedral clusters of the amorphous $Al_{75}Cu_{15}V_{10}$ alloy. In other words, Cu atoms share both the Al and V sites in the icosahedral clusters. Similar behavior is also observed in the icosahedral phase. All peaks in the intensity difference of the icosahedral phase in Fig. 6.21 are positive and similar to those of the original intensity profile in the quasi-crystalline state. Thus, the same concluding remark obtained for the amorphous state is also suggested in the icosahedral case. However, the following point is stressed with respect to the intensity difference, which provides the environmental structure around Cu, by comparing it with the original intensity peaks of 100000 and 110000: These peak intensities are reversed in the differential intensity profile, which implies the possibility that the Cu atoms only occupy particular sites of Al or V atoms in the icosahedral clusters in the quasi-crystalline state. Such selective replacement with the Cu atoms at Al and V sites may be the origin for the very narrow composition range for the formation of a single icosahedral phase in the Al–Cu–V ternary system. The environmental RDFs around Cu for both the samples are given in Fig. 6.22 together with the ordinary RDFs for comparison [41]. It is worth mentioning that the absolute values of the ordinary and environmental RDFs indicate the different physical meaning, and thus only the peak positions provide significant information in this comparison. As easily seen in the results of Fig. 6.22, peak maxima of oscillations in the ordinary and environmental RDFs are surprisingly coincident with each other in both the amorphous and icosahedral phases. This clearly implies that there is no significant difference in the atomic configuration around Cu from the average atomic structure; thus, it is plausible that the atomic ordering close to the icosahedral clusters in the amorphous phase grows preferentially to form a quasi-crystalline structure. However, the present results are not enough to provide an understanding of the mechanism of structural evolution from an amorphous matrix in this particular alloy.

A thermodynamically stable single icosahedral phase of the $Au_{65}Cu_{20}TM_{15}$ (TM = Fe, Ru and Os) alloy was found to be formed by casting and was fully annealed at about 1000 K [52]; its sufficiently higher degree of structural perfection has also been suggested [53]. Thus, these quasi-crystalline alloys appear to be one of the most appropriate systems to investigate structural details. The essential points of the results for these particular alloys obtained by the AXS method are described below, using the results of the icosahedral $Au_{65}Cu_{20}Fe_{15}$ alloy as an example.

The scattering intensities of the as-quenched $Au_{65}Cu_{20}Fe_{15}$ alloy measured at 6.811 and 7.086 keV in the close vicinity of the Fe-K absorption edge (7.112 keV) and their difference are given in Fig. 6.23. The peaks were indexed using the indexing scheme for icosahedral reflections proposed by Elser [54] and Cahn et al. [55]. The numbers in parentheses at some peaks denote the directions of the symmetrical axes along which the reflections are located. The profile of the intensity difference provided at the top of

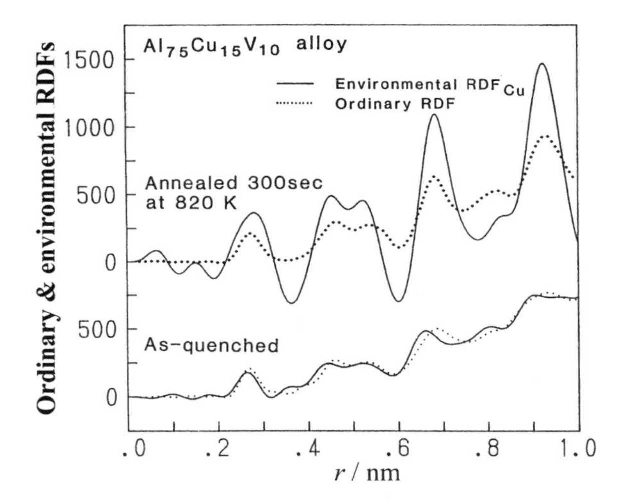


**Fig. 6.22.** Comparison of the environmental RDF around Cu of the amorphous $Al_{75}Cu_{15}V_{10}$ alloy with that of the icosahedral case. The *dotted lines* denote the ordinary RDFs [41]

Fig. 6.23 is essentially similar to the original intensity profiles, except for the small difference in relative peak intensity. It implies that the Fe atoms are homogeneously distributed in the icosahedral phase of this alloy.

On the other hand, similar AXS measurements for the same sample were carried out at energies of 8.680 and 8.955 keV below the Cu-K absorption edge (8.979 keV), and the results are given in Fig. 6.24. In contrast to the results of Fig. 6.23 for the environmental structure around Fe, the differential intensity profile appears to be significantly different from the original ones. The differential peak intensities indicate both positive and negative values and a change in each symmetry axis. Thus, it may be concluded that the Cu atoms are rather orderly arranged at particular sites in the icosahedral structure and not distributed uniformly along every symmetrical direction like Fe. In other words, the AXS results suggest that the directional dependence along the symmetry axes is strong for Cu atoms but not for Fe atoms.

The environmental RDFs around Fe and Cu are given in Fig. 6.25 together with the ordinary RDF. In the ordinary RDF, the first peak at 0.26 nm has a shoulder at the longer-distance side. Such a right-skewed first peak was explained by the presence of icosahedrally ordered clusters. On the other hand, both of the first peaks in the environmental RDFs for Fe and Cu have no shoulder, and their peak position coincides with the position of the first peak in the ordinary RDF. Furthermore, no clear peak nor shoulder are observed at the position of its shoulder. Taking account of the definition of the environmental RDFs and the concentrations of the constituents, the first peak can be attributed to the correlation of Al–Fe and Al–Cu pairs and its

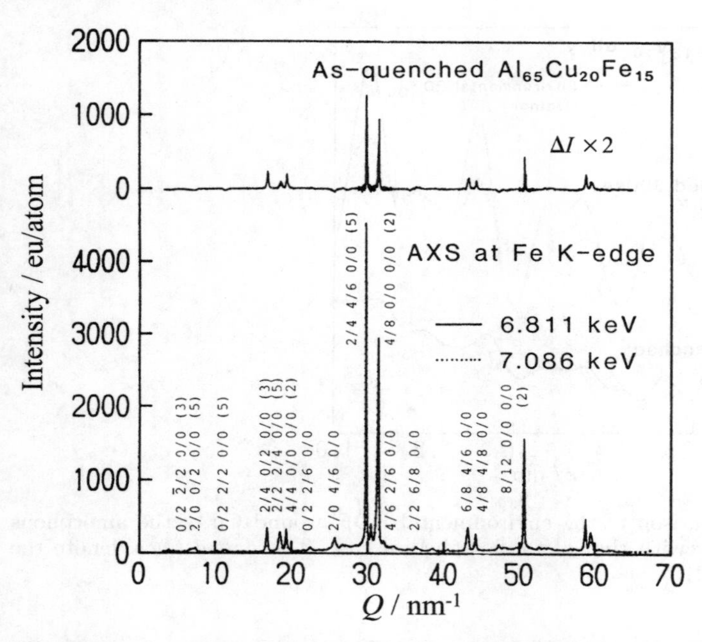


**Fig. 6.23.** Intensity profiles of the as-quenched $Al_{65}Cu_{20}Fe_{15}$ alloy obtained from AXS measurements at the Fe-K absorption edge [43]. *Top*: intensity difference

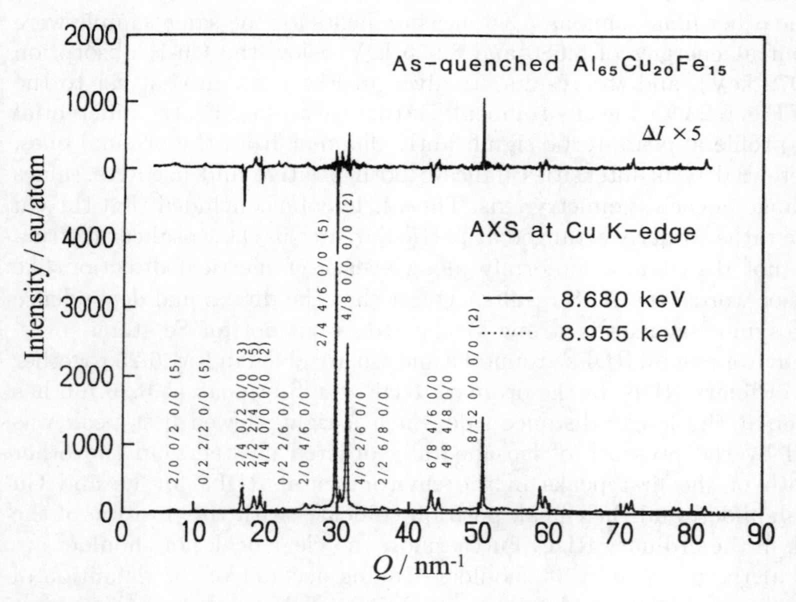


**Fig. 6.24.** Intensity profiles of the as-quenched $Al_{65}Cu_{20}Fe_{15}$ alloy obtained from AXS measurements at the Cu-K absorption edge [43]. *Top*: intensity difference

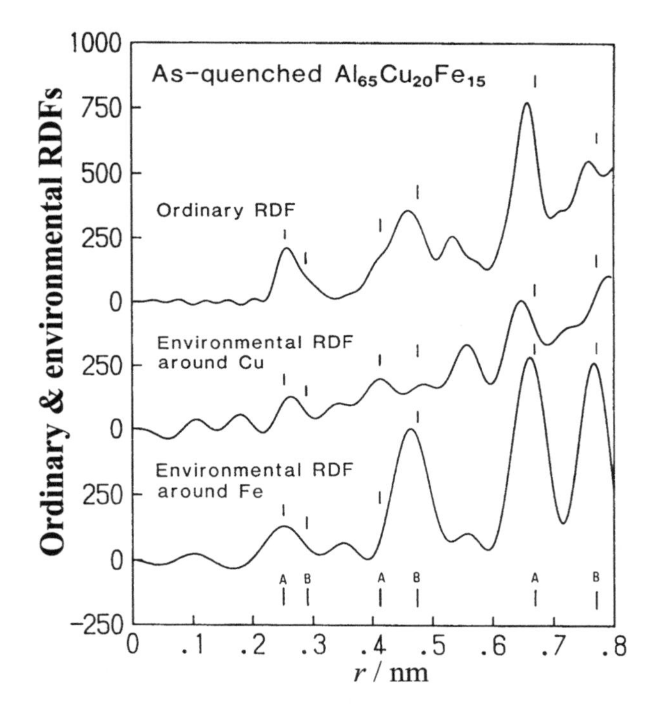


**Fig. 6.25.** Ordinary and environmental RDFs around Fe and Cu of the as-quenched $Al_{75}Cu_{15}Fe_{10}$ alloy [43]

shoulder to that of Al–Al pairs. A slight variation between the positions of the first peak in both of the environmental RDFs can be qualitatively explained by the different sizes of the Fe and Cu atoms. Some positions, which are $n\tau$ times the atomic distances of the first peak and its shoulder, are indicated by the arrows labeled "A" and "B" in Fig. 6.25, where $n$ is an integer and $\tau$ is the golden mean. Most of the peaks or shoulders in the RDFs are found to be in accord with these specific positions. This coincidence supports the results found by high-resolution electron microscopy that a scaling rule for $\tau$ exists in real space in the icosahedral $Al_{65}Cu_{20}Fe_{15}$ alloy [56].

The AXS method was applied to the decagonal phase, which is a two-dimensional quasi-crystal [42, 46]. The AXS profiles of the as-quenched decagonal $Al_{75}Fe_{15}Ni_{10}$ alloy measured at two energies below the Fe-K absorption edge are shown in Fig. 6.26. Similarly, Fig. 6.27 shows the intensity profiles obtained at the two energies of 8.306 and 8.031 keV below the Ni-K absorption edge (8.332 keV) and their difference. The intensity differences provided at the top of these two figures resemble the original intensity profiles. These AXS results suggest that both Fe and Ni atoms are homogeneously distributed and probably occupy similar atomic sites in the decagonal structure

**Table 6.6.** Coordination numbers and distances of nearest-neighbor pairs in the as-quenched decagonal $Al_{75}Fe_{15}Ni_{10}$ alloy and the $Al_{13}Fe_4$ crystal

| | $r$ (nm) | $N$ |
|---|---|---|
| Environmental RDF for Fe | | |
| Fe–Al | 0.252 ± 0.002 | 7.6 ± 0.2 |
| Environmental RDF for Ni | | |
| Ni–Al | 0.270 ± 0.002 | 8.5 ± 0.3 |
| Ordinary RDF | | |
| Al–Al | 0.284 ± 0.002 | 8.6 ± 1.1 |
| $Al_{13}Fe_4$ crystal | | |
| Fe–Al | 0.254 | 9.7 |
| Al–Al | 0.279 | 7.6 |

of the $Al_{75}Fe_{15}Ni_{10}$ alloy. Such feature might be closely related to the experimental result that a single decagonal phase can be formed in a relatively wide composition range, e.g. 9 to 16 at.-% Ni and 9 to 21at.-% Fe in the ternary Al–Fe–Ni alloy [33].

Figure 6.28 shows the ordinary RDF and environmental RDFs for Fe and Ni in the decagonal $Al_{75}Fe_{15}Ni_{10}$ alloy [42]. As seen in Fig 6.28, the essential profiles of the three RDFs are basically identical. However, the following points are suggested: The first peak denoted by "a" in the ordinary RDF has a shoulder labeled as "b", while the first peaks in the environmental RDFs for Fe and Ni have no shoulder. Thus, the first peak observed at "a" is attributed to Al–Fe and Al–Ni pairs, and its shoulder at "b" to Al–Al pairs. Coordination numbers and atomic distances estimated from the three sets of RDF data are summarized in Table 6.6. Since the Al–Fe and Al–Ni pairs are located at the maximum of the first peak in the ordinary RDF, the tabulated coordination number is computed as an average value for the Al atoms around Fe and Ni.

It is worth mentioning that the distances of the Al–Fe and Al–Al pairs estimated from the RDFs are very close to the nearest-neighbor distances of the Al–Fe and Al–Al pairs in $Al_{13}Fe_4$ [57] and their coordination numbers are roughly equal to the values of $Al_{13}Fe_4$. Henley [58] demonstrated that the layer structure of the crystalline $Al_{13}Fe_4$ alloy can be decomposed into the rhombic tiles of the two-dimensional Penrose tiling. Figure 6.29 illustrates the puckered layer structure which is one of the two layers of the $Al_{13}Fe_4$ structure [57] with a decoration of the rhombic tiles [58]. The distances labeled "a", "b", "d" and "e" in this figure correspond to the distances given in Fig. 6.28. This suggests that the local atomic structure of the decagonal $Al_{75}Fe_{35}Ni_{10}$ alloy is very similar to that found in the $Al_{13}Fe_4$ crystal. Since the atomic distances between Al and Fe atoms, or Fe and Fe atoms belonging to different layers along the $c$-axis in the $Al_{13}Fe_4$ crystal, are approximately

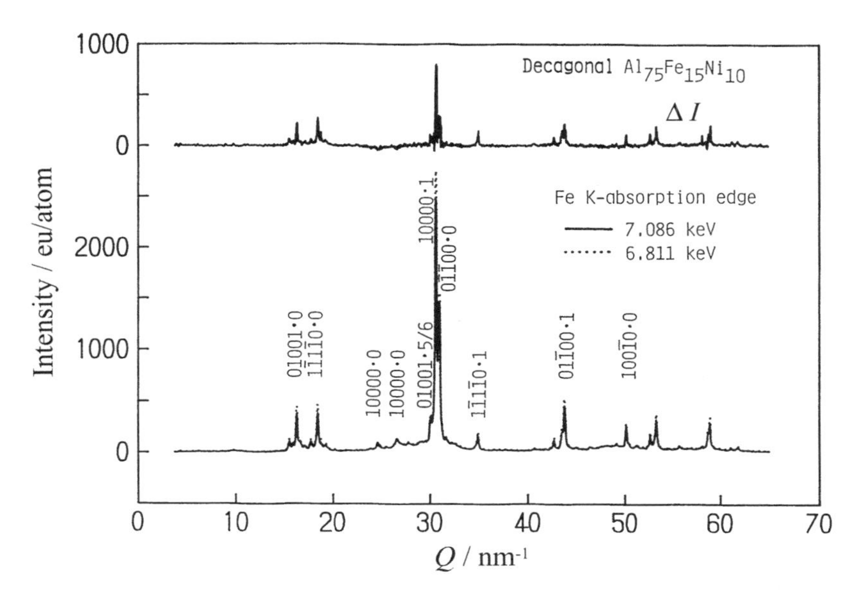


**Fig. 6.26.** Intensity profiles of the as-quenched decagonal $Al_{75}Fe_{15}Ni_{10}$ alloy obtained from AXS measurements at the Fe-K absorption edge [42]. *Top*: intensity difference

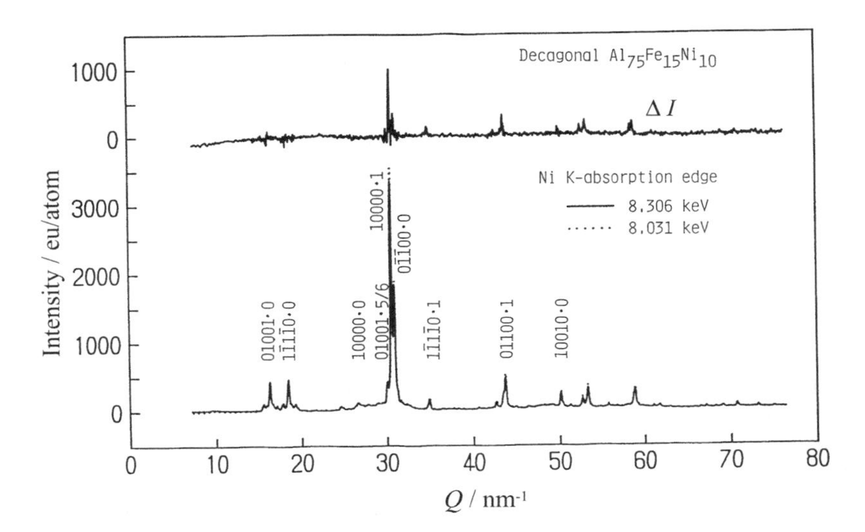


**Fig. 6.27.** Intensity profiles of the as-quenched decagonal $Al_{75}Fe_{15}Ni_{10}$ alloy obtained from AXS measurements at the Ni-K absorption edge [42]. *Top*: intensity difference

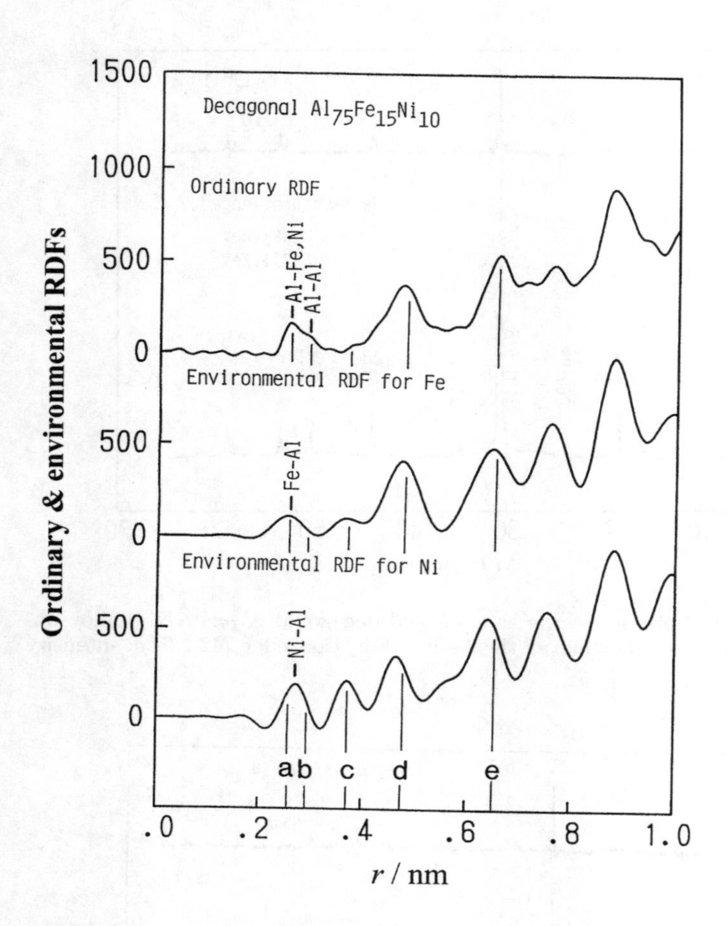


**Fig. 6.28.** Ordinary and environmental RDFs around Fe and Ni of the as-quenched decagonal $Al_{75}Fe_{15}Ni_{10}$ alloy [42]

equal to the distance "c"; the peak labeled by "c" may be attributed to some atomic arrangements perpendicular to the decagonal plane. It should also be mentioned that the local atomic structure of some icosahedral or decagonal phases is successfully approximated by a particular crystalline structure with a very large unit cell [59]. The result obtained using the AXS method proved the similarity of the local atomic structure between the decagonal $Al_{75}Fe_{35}Ni_{10}$ alloy and the $Al_{13}Fe_4$ crystal.

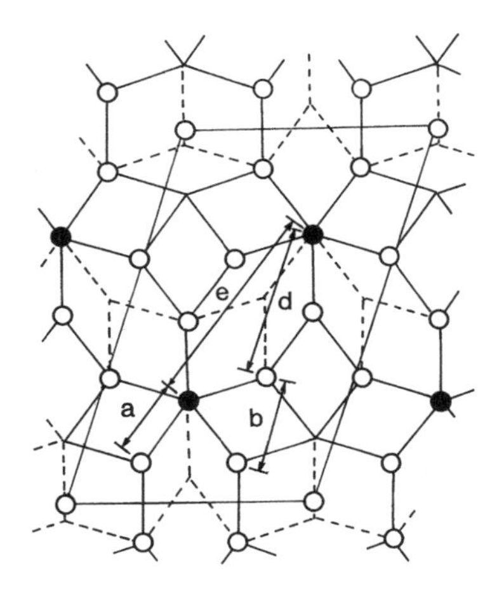


**Fig. 6.29.** The puckered layer of the $Al_{13}Fe_4$ structure (○: Al; ●: Fe) with the decoration of the two-dimensional Penrose tiling proposed by Henley [58]

## References


1. H.M. Rietveld: J.Appl. Crystallogr., **2**, 65 (1969)
2. D. Shechtman, I.A. Blech, D. Gratias and J.W. Cahn: Phys. Rev. Lett., **53**, 1951 (1984)
3. X. Zhu, R. Birringer, U. Herr and H. Gleiter: Phys. Rev., B **35**, 9085 (1987)
4. J.G. Bednorz and K.A. Müller: Z. Phys., **64**, 189 (1986)
5. R.J. Hill, J.R. Craig, and G.V. Gibbs: Phys.Chem.Miner., **4**, 317 (1979)
6. R.W. James: *The Optical Principles of the Diffraction of X-rays* (G.Bells, London 1954)
7. G. Metrlik, C.J. Sparks and K. Fischer (editors): *Resonant Anomalous X-ray Scattering* (North-Holland Amsterdam 1994)
8. K. Shinoda, K. Sugiyama, C. Reynales, Y. Waseda and K.T. Jacob: Shigen-to-Sozai, **111**, 801 (1995)
9. Y. Waseda and M. Saito: J. Jpn. Soc. Synchrotron Radiat. Res., **10**, 299 (1997)
10. Y. Waseda, K. Shinoda and K. Sugiyama: Z. Naturforsch., **50a**, 1199 (1995)
11. H. Toraya, and F. Marumo: Rep. Res. Lab. Eng. Mater., Tokyo Inst. of Tech., **5**, 55 (1980)
12. H. Toraya: J.Appl.Crystallogr., **19**, 440 (1986)
13. H.St.C. O'Neill and A. Navrotsky: Am.Miner., **68**, 181 (1983)
14. G.A. Sawatzky, F. Van der Wande, A.H. Mirrush: Phys. Rev, **187**, 747 (1969)
15. K. Shinoda, K. Sugiyama, K. Omote and Y. Waseda: Int. J. Soc. Mater. Eng. Resour., **4**, 20 (1996)
16. C.W. Chu, P.H. Hor, R.L. Meng, L. Gao, Z.J. Huang and Y.Q. Wang: Phys. Rev. Lett., **58**, 405 (1987)
17. *Special Suppl. Issue of Advanced Ceramic Materials* Am. Ceram. Soc., **2** (1987) No. 3B
18. K. Sugiyama and Y. Waseda: Mater. Trans. JIM, **30**, 235 (1989)
19. F. Izumi, H. Asano, T. Ishigaki, A. Ono and F.P. Okumura: Jpn. J. Appl. Phys., **26**, L611 (1987)
20. K. Kitazawa, K. Kishio, H. Takagi, T. Hasegawa, S. Kanbe, S.Uchida, S. Tanaka and K. Fueki: Jpn. J. Appl. Phys., **26**, L339 (1987)
21. For example, F.Izumi: MAC Sci. J. **2**, 2 (1988)

22. P. Rahlfs: Z. Phys. Chem., B **31**, 157 (1936)
23. T. Sakuma, K. Sugiyama, E. Matsubara and Y. Waseda: Mater. Trans. JIM, **30**, 365 (1989)
24. E. Matsubara, K. Okuda, Y. Waseda and T. Saito: Z. Naturforsch., **47a**, 1023 (1992)
25. W.H. Hall: Proc. Phys. Soc. Lond. A **62**, 741 (1949)
26. T. Sato, K. Haneda, M. Seki and T. Iijima: *Proc. Int. Symp. Physics of Magnetic Materials* (World Scientific, Singapore 1987), pp. 210
27. B.E. Warren: *X-ray Diffraction* (Addison-Wesley, Reading 1969)
28. F. Betts and A. Bienenstock: J. Appl. Phys.,**143**, 4591 (1972)
29. E. Matsubara and Y. Waseda in: *Resonant Anomalous X-ray Scattering–Theory and Application*, ed. by G. Materlik, C.J. Sparks and K. Fischer (North-Holland, Amsterdam 1994) pp. 345
30. E. Matsubara and Y. Waseda: J. Phys. Condens. Matter, **1**, 8575 (1989)
31. K.S. Liang, S.S. Laderman and J.H. Sinfelt: J. Chem. Phys., **86**, 2352 (1987)
32. P.H. Fuoss, P. Eisengerger, W.K. Warburton and A. Bienenstock: Phys. Rev. Lett., **46**, 1537 (1981)
33. A.P. Tsai, A. Inoue and T. Masumoto: Mater. Trans. JIM, **30**, 150 (1989)
34. A.P. Tsai, A. Inoue and T. Masumoto: Mater. Trans. JIM, **30**, 300 (1989)
35. P.J. Steihardt, D.R. Nelson and M. Ronchetti: Phys. Rev., B **28**, 784 (1983)
36. S. Sachdev and D.R. Nelson: Phys. Rev. Lett., **53**, 1947 (1984)
37. C.L. Henley: Commun. Condens. Matter Phys., **13**, 59 (1987)
38. D.D. Kofalt, S. Nanao, T. Egami, K.M. Wong and S.J. Poon: Phys. Rev. Lett., **57**, 114 (1986)
39. D.D. Kofalt, I.A. Morrison, T. Egami, S. Priesche, S.J. Poon and P.J. Steinhardt: Phys. Rev., B **35**, 4489 (1987)
40. R. Fuchs, S.B. Jast, H.J. Güntherodt and P. Fischer: Z. Phys., B **68**, 309 (1987)
41. E. Matusbara, Y. Waseda, A.P. Tsai, A. Inoue and T. Masumoto: J. Mater. Sci., **25**, 2507 (1990)
42. E. Matusbara, Y. Waseda, A.P. Tsai, A. Inoue and T. Masumoto: Z. Naturforsch., **46a**, 605 (1991)
43. E. Matusbara, Y. Waseda, A.P. Tsai, A. Inoue and T. Masumoto: Z. Naturforsch., **45a**, 50 (1990)
44. E. Matusbara and Y. Waseda: *Proc. China–Japan Seminars on Quasi-crystals*, ed. by K.H. Kuo and T. Ninomiya (World Scientific, Singapore 1991) pp. 96
45. R. Hu, T. Egami, A.P. Tsai, A. Inoue and T. Masumoto: Phys. Rev., B **46**, 6105 (1992)
46. E. Matusbara, Y. Waseda, A.P. Tsai, A. Inoue and T. Masumoto: Mater. Trans. JIM, **34**, 151 (1993)
47. E. Matusbara, K. Harada, Y. Waseda, H.S. Chen, A. Inoue and T. Masumoto: J. Mater. Sci., **23**, 753 (1988)
48. S. Garcon, P. Sainfort, G. Regazzoni and J.M. Duboir: Script Metall., **21**, 1493 (1987)
49. P.J. Brown: Acta Crystallogr., **12**, 995 (1959)
50. Y. Waseda: *The Structure of Non-Crystalline Materials* (McGraw-Hill, New York 1980)
51. P.A. Bancel, P.A. Heiney, P.W. Stephens, A.I. Goldman and P.M. Horn: Phys. Rev. Lett., **54**, 2422 (1985)
52. A.P. Tsai, A. Inoue and T. Masumot: Jpn. J. Appl. Phys., **26**, L1505 (1987)
53. C.A. Guryan, A.I. Goldman, P.W. Stephens, K. Hiraga, A.P. Tsai, A. Inoue and T. Masumoto: Phys. Rev. Lett., **62**, 2409 (1989)
54. V. Elser: Acta Crystallogr., A **42**, 36 (1986)
55. J.W. Cahn, D. Schechtman and D. Gratias: J. Mater. Res., **1**, 13 (1986)

56. K. Hiraga, B.P. Zhang, M. Hirabayashi, A. Inoue and T. Masumoto: Jpn. J Appl. Phys., **27**, L951 (1988)
57. P.J. Black: Acta Crystallogr., **8**, 175 (1955)
58. C.L. Henley: J. Non-Cryst. Solids, **75**, 91 (1985)
59. K. Edagawa, K. Suzuki, M. Ichihara, S. Takeuchi and T. Shibuya: Philos. Mag., B **64**, 629 (1991) 2-16

# 7. Selected Examples of Structural Determination for Non-crystalline Materials Using the AXS Method

A simple and quantitative description with a few parameters of distance and angle, as employed for crystalline materials, is impossible for melt and glass structures, mainly because of the particular non-periodicity in their atomic arrangements which results in the fluctuation of both the atomic position and angle. In addition, non-crystalline materials of interest mostly contain more than two kinds of elements, and therefore the concept and utility of the partial structure functions in multi-component, non-crystalline materials have been emphasized for a long time. We can have more than just one-dimensional radial distribution function (RDF) information, when the full set of the partial functions are obtained. Three partial functions are required in a binary system, and six partials in a ternary system. However, the actual implementation of this subject is not a trivial task even for binary systems.

The partial structure factors, corresponding to the Fourier transform of RDFs, for a binary system can be estimated only by making available at least three independent intensity measurements, for which the weighting factors are varied without any change in their RDFs. For this purpose, the isotope substitution method for neutrons, firstly applied by Enderby et al. [1] to liquid $Cu_6Sn_5$ alloys and to various molten salts [2,3,4,5,6,7,8,9,10,11,12,13,14,15], is considered undoubtedly one of the powerful methods. However, it seems intrinsically to be somewhat limited in practice by the availability of suitable isotopes, and the structure is automatically assumed to remain identical upon substitution by the isotopes.

On the other hand, following the pioneer work of Bondot in 1974 for $GeO_2$ glass [16] estimating the near-neighbor correlation from X-ray diffraction data, where the scattering ability is varied by the anomalous dispersion factors of Ge ($f' = -1.31$, $f'' = 1.04$ for Cu-K$\alpha$ and $f' = 0.38$, $f'' = 1.29$ for Ag-K$\alpha$), a number of developments have been devoted to the anomalous X-ray scattering (AXS) method for determining the partial functions in various binary systems. Our present understanding of the AXS method for this subject is still far from complete, but its principle should basically work well without any assumptions for many more elements in the periodic table, and its potential power, in the author's view, may not be overemphasized. For this reason, the current experimental results on partial functions are given below with some selected examples.

## 7.1 Partial Structure Functions in Molten Salts

The individual partial structure functions of some molten salts such as CuBr [17] and RbBr [18] were recently estimated from the AXS measurements coupled with a devised simulation technique. The availability of a synchrotron-radiation source has dramatically improved both the acquisition and quality of AXS data by enabling the use of an energy in which AXS is maximized. However, even when the anomalous dispersion term can be varied in the close vicinity of the absorption edge, the so-called determinant of the normalized weighting matrix [3] is often very small, making the accuracy of the partial functions rather poor [19, 14]. This means the insufficient absolute accuracy of the AXS experimental data still prevents us from obtaining exact solutions to the simultaneous linear equations from the AXS measurements alone. The reverse Monte Carlo (hereafter referred to as RMC) simulation technique, originally proposed by McGreevy and his colleague [20, 21, 22, 23], appears to provide one useful way of reducing such inconvenience. This data processing for AXS with the help of the RMC technique is not covered in the previous leading work on molten $GeBr_4$ [14].

Figure 7.1 shows the coherent scattering intensity profiles of molten RbBr in electron units per atom obtained from the measurements at incident energies of 13.170, 13.445, 14.902 and 15.177 keV. These four energies correspond to 300 and 25 eV below the K absorption edges of Br (13.470 keV) and Rb (15.202 keV), respectively [18]. The intensity profile measured at an incident energy of 17.0 keV, which is considerably far away from both absorption edges, is also given in this figure. The resultant scattering intensities appear to significantly depend upon the incident X-ray energy, and these variations in intenstity should be attributed to the change in the anomalous dispersion terms of Rb and Br. The anomalous dispersion terms for Rb and Br for these energies are listed in Table 7.1. The normal atomic scattering factors are taken from the *International Tables for X-ray Crystallography.*

**Table 7.1.** Anomalous dispersion terms for Cu, Rb and Br

| Energy (keV) | Cu | | Rb | | Br | |
|---|---|---|---|---|---|---|
| 8.680 | $f' = -3.06$ | $f'' = 0.51$ | | | $f' = -0.90$ | $f'' = 1.12$ |
| 8.955 | $f' = -5.58$ | $f'' = 0.49$ | | | $f' = -0.96$ | $f'' = 1.06$ |
| 13.170 | $f' = 0.00$ | $f'' = 2.06$ | $f' = -1.79$ | $f'' = 0.66$ | $f' = -3.37$ | $f'' = 0.53$ |
| 13.445 | $f' = 0.04$ | $f'' = 1.99$ | $f' = -1.90$ | $f'' = 0.64$ | $f' = -5.82$ | $f'' = 0.51$ |
| 14.902 | | | $f' = -3.47$ | $f'' = 0.53$ | $f' = -1.21$ | $f'' = 3.19$ |
| 15.177 | | | $f' = -5.91$ | $f'' = 0.51$ | $f' = -1.40$ | $f'' = 3.10$ |
| 17.000 | $f' = 0.26$ | $f'' = 1.33$ | $f' = -1.29$ | $f'' = 3.11$ | $f' = -0.48$ | $f'' = 2.57$ |

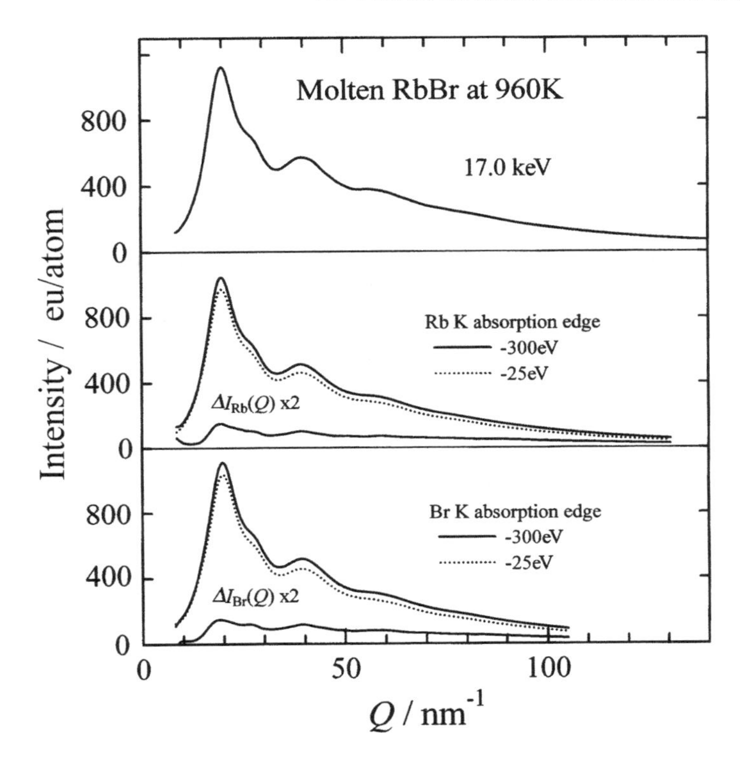


**Fig. 7.1.** Coherent intensity profiles of molten RbBr at 960 K measured at incident energies of 13.170, 13.445, 14.902, 15.117 and 17.0 keV

It may be helpful to recall the essential points of the AXS method. When the incident energy is set close below the absorption edge, $E_{\mathrm{abs}}$, of a specific element, for example, Rb in the present case, the anomalous dispersion phenomena becomes significant, and then the variation between the two intensities, $\Delta i_{\mathrm{Rb}}(Q, E_1, E_2)$, measured at incident energies of $E_1$ and $E_2$, is attributed to a change in the real part of anomalous dispersion term of Rb. Therefore, the following relation can readily be obtained [18]:

$$
\begin{aligned}
\Delta i_{\mathrm{Rb}}(Q, E_1, E_2) &\equiv \frac{I'(E_1) - I'(E_2)}{c_{\mathrm{Rb}}\left\{f'_{\mathrm{Rb}}(E_1) - f'_{\mathrm{Rb}}(E_2)\right\}W(Q, E_1, E_2)} \\
&= \frac{c_{\mathrm{Rb}}\Re\left\{f_{\mathrm{Rb}}(Q, E_1) + f_{\mathrm{Rb}}(Q, E_2)\right\}}{W(Q, E_1, E_2)}[a_{\mathrm{RbRb}}(Q) - 1] \\
&+ \frac{c_{\mathrm{Br}}\Re\left\{f_{\mathrm{Br}}(Q, E_1) + f_{\mathrm{Br}}(Q, E_2)\right\}}{W(Q, E_1, E_2)}[a_{\mathrm{RbBr}}(Q) - 1], \quad (7.1)
\end{aligned}
$$

$$I'(E_k) = I(Q, E_k) - \langle f^2(Q, E_k) \rangle \qquad (k = 1 \text{ or } 2), \tag{7.2}$$

$$W(Q, E_1, E_2) = \sum_{k=1}^{2} c_k \Re\Big\{ f_k(Q, E_1) + f_k(Q, E_2) \Big\}, \tag{7.3}$$

where $E_1 < E_2 < E_{\rm abs}$ and $\Re$ denotes the real part of the values in the brackets. Then, the quantity $\Delta i_{\rm Rb}(Q, E_1, E_2)$ associated with Rb contains two partial structure factors, $a_{\rm RbRb}(Q)$ and $a_{\rm RbBr}(Q)$. Similarly, $\Delta i_{\rm Br}(Q, E_1, E_2)$ measured at the lower-energy side of the Br-K absorption edge involves $a_{\rm BrBr}(Q)$ and $a_{\rm RbBr}(Q)$. Thus, the set of equations to be solved for the partial structure factors is expressed in the following generalized form:

$$\Big[I\Big] = \Big[M\Big] \cdot \Big[S\Big], \tag{7.4}$$

$$[I] = \begin{pmatrix} \Delta i_{\rm A}(Q, E_1, E_2) \\ \Delta i_{\rm B}(Q, E_3, E_4) \\ i(Q, E_5) \end{pmatrix}, \qquad [S] = \begin{pmatrix} a_{\rm AA}(Q) - 1 \\ a_{\rm AB}(Q) - 1 \\ a_{\rm BB}(Q) - 1 \end{pmatrix}, \tag{7.5}$$

$$[M] = \begin{bmatrix} w_{11} & w_{12} & w_{13} \\ w_{21} & w_{22} & w_{23} \\ w_{31} & w_{32} & w_{33} \end{bmatrix}, \tag{7.6}$$

$$\left.\begin{array}{ll} w_{11} = \frac{c_{\rm A}\Re\{f_{\rm A}(Q,E_1)+f_{\rm A}(Q,E_2)\}}{W(Q,E_1,E_2)} & w_{12} = \frac{c_{\rm B}\Re\{f_{\rm B}(Q,E_1)+f_{\rm B}(Q,E_2)\}}{W(Q,E,E_2)} \\ w_{22} = \frac{c_{\rm A}\Re\{f_{\rm A}(Q,E_3)+f_{\rm A}(Q,E_4)\}}{W(Q,E_3,E_4)} & w_{23} = \frac{c_{\rm B}\Re\{f_{\rm B}(Q,E_3)+f_{\rm B}(Q,E_4)\}}{W(Q,E_3,E_4)} \\ w_{31} = \frac{c_{\rm A}^2 f_{\rm A}^2(Q,E_5)}{\langle f \rangle^2} & w_{32} = \frac{2c_{\rm A}c_{\rm B}f_{\rm A}(Q,E_5)f_{\rm B}(Q,E_5)}{\langle f \rangle^2} \\ w_{33} = \frac{c_{\rm B}^2 f_{\rm B}^2(Q,E_5)}{\langle f \rangle^2} & \\ w_{13} = w_{21} = 0 & \end{array}\right\} . \tag{7.7}$$

In these equations, the subscripts A and B correspond to Rb and Br, respectively.

The environmental interference functions of molten RbBr at 960 K, $Q\Delta i_{\rm Br}(Q)$ for Br and $Q\Delta i_{\rm Rb}(Q)$ for Rb, calculated by taking the difference between the two intensities using (7.1), are given in Fig. 7.2, together with the ordinary interference function, $Qi(Q)$.

Figure 7.3 provides three partial structure factors of molten RbBr estimated directly from three experimental data sets of five different energies of Fig. 7.1 by solving the simultaneous linear equation (7.4). It is found that the results are rather widely spread in certain positions. Namely, the numerical solutions of (7.4) obtained directly from three experimental data sets appear to be ill conditioned in several $Q$ values [3, 24]. This is mainly due to the experimental uncertainty such as the relatively small difference between the anomalous dispersion terms at two energies and the unpredictable large fluctuation in the numerical solution of (7.4), when the pivot of the matrix is

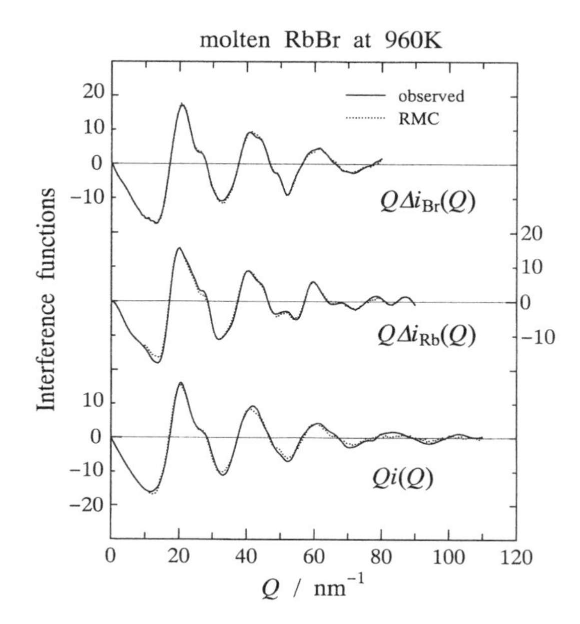


**Fig. 7.2.** The environmental interference functions $Q\Delta i_{\mathrm{Rb}}(Q)$ for Rb and $Q\Delta i_{\mathrm{Br}}(Q)$ for Br in molten RbBr at 960 K, together with the ordinary interference function, $Qi(Q)$. *Solid lines* correspond to the experimental data. *Dotted lines* denote values calculated by the RMC simulation technique

close to zero [25,26]. Nevertheless, the author maintains the opinion that the variations detected are sufficiently sizable. One can easily find, at this stage, that the first trough of unlike-atom pair $a_{\mathrm{RbBr}}(Q)$ is situated at a $Q$ value where the principal peaks for like-atom pairs are detected, and the basic profile of $a_{\mathrm{RbRb}}(Q)$ is similar to that of $a_{\mathrm{BrBr}}(Q)$. Such structural features are very similar to those of simple ionic liquids such as molten alkali halide [3,6].

In Fig. 7.3, the poor conditioning of simultaneous linear equations certainly appears in several $Q$ values. Small experimental errors in the total scattering intensities are known to induce unpredictable large fluctuation in the partial structure factors, and some physically unreasonable behavior frequently appears in the resultant RDFs, as already reported in a number of works [3,12,15,24]. Such small experimental errors cannot always be avoided to the best knowledge of the author. Therefore, some reservations should be stressed regarding the quantitative accuracy of the partial functions obtained from the AXS data, although some sophisticated methods of data processing for AXS have been proposed and examined [19,14,27]. This is particularly true when the numerical solutions of (7.4) are considered, from a standard

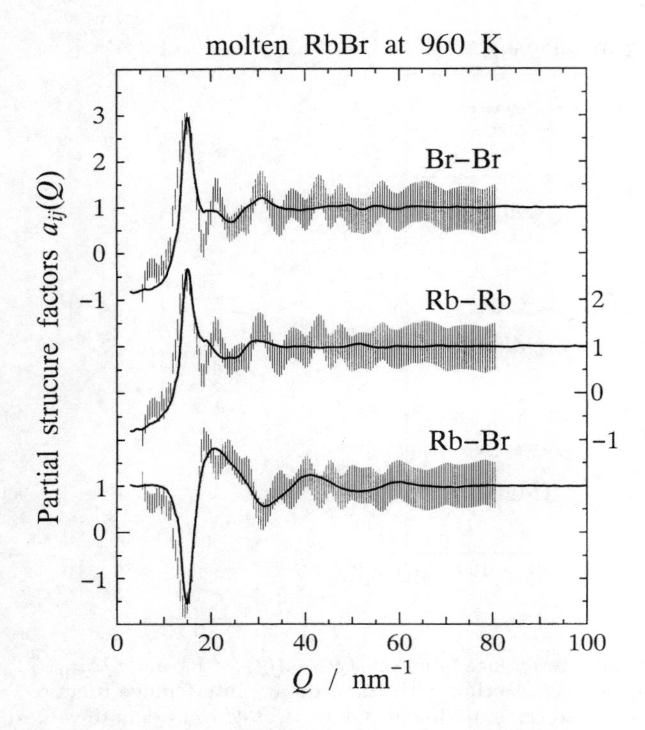


**Fig. 7.3.** Three partial structure factors for molten RbBr at 960 K. *Solid lines* correspond to the values calculated by the RMC simulation technique

mathematical point of view. In regard to this subject, the author rather takes the view that we could obtain the range of partial functions so as to reproduce the experimental data and to satisfy known physical constraints, as shown in the first work on the partials of liquid $Cu_6Sn_5$ by Enderby et al. [1]. This opinion is also based on the following reason:The normalized determinant suggested by, for example, Edwards et al. [3] is one way of indicating the conditioning of the matrix expressed by weighting factors in (7.4). For a reliable separation of the partials, the normalized determinant should be as near to unity as possible. The normalized determinant is defined as follows:

$$|M|_{\mathrm{n}} = \begin{vmatrix} w_{11}/\overline{w_1} & w_{12}/\overline{w_1} & w_{13}/\overline{w_1} \\ w_{21}/\overline{w_2} & w_{22}/\overline{w_2} & w_{23}/\overline{w_2} \\ w_{31}/\overline{w_3} & w_{32}/\overline{w_3} & w_{33}/\overline{w_3} \end{vmatrix} \left( \overline{w_i} = \sqrt{\sum_{j=1}^{3} w_{ij}^2} \right) . \tag{7.8}$$

In order to grasp the conditioning of recent AXS measurements, the normalized determinant of some binary liquids are plotted in Fig. 7.4 as a function of wave vector. Some typical values obtained in the isotope substitution method for neutrons are also given in this figure for comparison. For example, the nor-

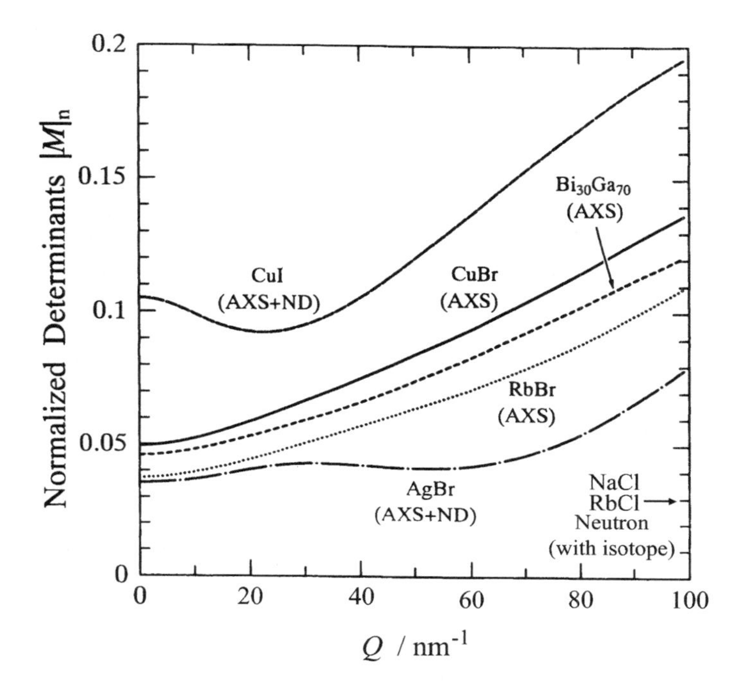


**Fig. 7.4.** Examples of the normalized determinant, defined by Edwards et al. [3], for some binary liquids as a function of wave vector

malized determinant is 0.037 for RbBr and 0.049 for CuBr at $Q = 0.0$ nm$^{-1}$ and 0.063 for RbBr and 0.084 for CuBr at $Q = 50$ nm$^{-1}$; similar numerical values are recognized at all $Q$ values. These conditions are comparable to, or slightly better than, the values of isotope substitution experiments for neutrons, in which they are usually of the order of 0.03 or less for molten salts. Thus, an improvement in reliability for recent AXS experiments, although it is still far from ideal, might be obtained by increasing the normalized determinant.

The use of the RMC simulation technique [20, 21, 22, 23] appears to hold promise in reducing the problem of the numerical solutions of (7.4) by choosing values with a reasonable physical meaning [17,18]. In the RMC simulation, the atomic configurations are estimated rather simply on the grounds of their consistency with the experimental data when comparing the case by using the pair potential. We start with an initial configuration of 4096 particles, half of them represent Rb, the remaining half Br, in a cubic box of size $L = 5.924$ nm. The usual periodic boundary conditions, where the cube is surrounded by images of itself, are employed, and the so-called partial pair distribution functions, $g_{ij}(r)$, are calculated. Then, the interference functions via structure factors computed by the Fourier transformation are compared with the experimental results by estimating the following statistic:

$$\chi^2 = \sum_{m=1}^{n} \frac{\{i(Q_m) - i^{\mathrm{c}}(Q_m)\}^2}{\sigma^2(Q_m)} + \sum_{\alpha}\sum_{m=1}^{n'} \frac{\{\Delta i_\alpha(Q_m) - \Delta i_\alpha^{\mathrm{c}}(Q_m)\}^2}{\sigma_\alpha^2(Q_m)}, \quad (7.9)$$

where $i^{\mathrm{c}}(Q_m)$ and $\Delta i_\alpha^{\mathrm{c}}(Q_m)$ are the calculated interference function and its difference for the $\alpha$ component, measured at $Q_m$, respectively. $\sigma(Q_m)$ and $\sigma_\alpha(Q_m)$ are the estimates of the experimental uncertainty. A new configuration is then generated by the random movement of one particle. It may be noted that the cut off distance is set for each $g_{ij}(r)$ to be closer than the direct contact value of two particles, in order to prevent particles coming un-physically close to one another. When the new configuration violates these cut off restrictions, it is rejected and the previous configuration data is restored. Otherwise the variation in $g_{ij}(r)$ is calculated, and from this new $i^{\mathrm{c}}(Q)$, $\Delta i^{\mathrm{c}}(Q)$ and $\chi^2$ are obtained. When the new value of $\chi^2$ is smaller than the old one, the new configuration is accepted; otherwise it is accepted only with a probability less than unity. This iteration process is carried out until $\chi^2$ indicates a reasonable convergence. The flow chart of such iterative procedure is given in Fig. 7.5.

The RMC simulation results are found to reproduce three independent interference functions: $Q\Delta i_{\mathrm{Rb}}(Q)$, $Q\Delta i_{\mathrm{Br}}(Q)$ and $Qi(Q)$, as shown in Fig. 7.2. This agreement clearly implies that the present approach basically works well. The solid lines in Fig. 7.3 are the RMC partial structure factors which are coincident with the average values of the experimental error bars, in order to draw a smooth continuous function. It should be kept in mind that the RMC simulation technique is not a unique mathematical procedure. However, the partial structure factors estimated from the present AXS measurements coupled with the RMC simulation technique are considered to be, at least, in a sense the necessary condition at best, although they might be not the sufficient condition.

Figure 7.6 provides the partial pair distribution functions, $g_{ij}(r)$, of molten RbBr. It can be found that $g_{\mathrm{RbRb}}(r)$ and $g_{\mathrm{BrBr}}(r)$ are approximately in anti-phase to $g_{\mathrm{RbBr}}(r)$. This indicates well-defined charge ordering. The

**Table 7.2.** Structural parameters in molten CuBr and RbBr ($r$: nm)

| | CuBr (810 K) | | | | RbBr (960 K) | | | | | |
|---|---|---|---|---|---|---|---|---|---|---|
| | Cu–Br | | Br–Br | | Rb–Rb | | Rb–Br | | Br–Br | |
| | $r_{+-}$ | $N$ | $r_{--}$ | $N$ | $r_{++}$ | $N$ | $r_{+-}$ | $N$ | $r_{--}$ | $N$ |
| Melt | 0.245 | 3.1 | 0.398 | 11.4 | 0.495 | 13.5 | 0.337 | 5.4 | 0.484 | 13.7 |
| Crystal | 0.247 | 4 | 0.403 | 12 | 0.501 | 12 | 0.354 | 6 | 0.501 | 12 |
| $r_-/r_+$ | 1.62 | | | | 1.44 | | | | | |

Experimental uncertainty: unlike-ion pair $r$: ±0.003; $N$: ±0.3
like-ion pair $r$: ±0.003; $N$: ±0.5

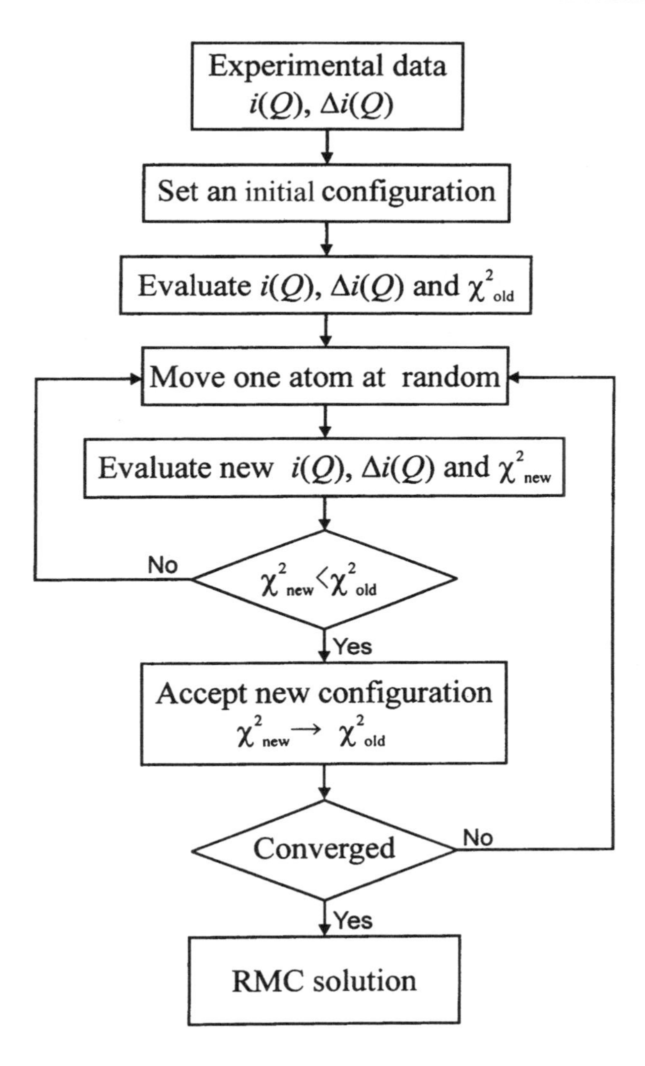


**Fig. 7.5.** Flow chart of the RMC simulation technique for analyzing the AXS data

peak positions and the coordination numbers estimated from these partial functions are summarized in Table 7.2, together with the corresponding values for the crystalline RbBr phase of the NaCl-type structure. The coordination numbers for unlike-ion pairs and like-ion pairs are estimated to be 5.4 and 13.5 on average, respectively. They are rather close to the crystalline values. It may also be worth mentioning that the ratio of nearest-neighbor peak positions, defined as $r_{--}/r_{+-}$, is 1.44, which is closer to the octahedron

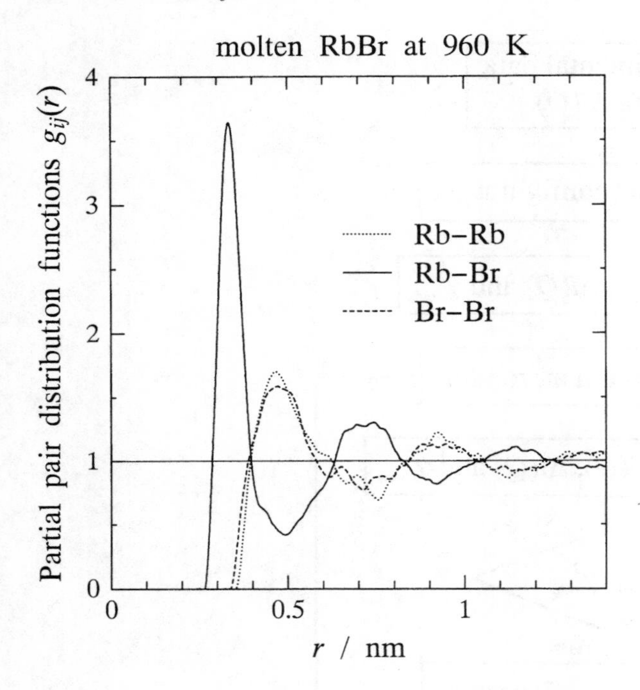


**Fig. 7.6.** Three partial pair distribution functions for molten RbBr at 960 K

value of 1.41 than the tetrahedron value of 1.63. Consequently, it is safe to suggest that the first coordination shell of the Rb–Br pair can be described by the octahedral configuration, as it is in the crystalline state and such structural features are quite similar to the case for molten alkali chlorides (see, for example, the studies of NaCl [3] and RbCl [6]). Figure 7.7 illustrates a representative ionic configuration generated in the RMC simulation process for molten RbBr. The $RbBr_6$ units are selected and drawn as octahedra in this figure. It can be seen that octahedral configurations for the nearest-neighbor Rb–Br pair are predominant in the molten state.

One additional example is given below, using the results of CuBr, which is considered to be a non-ideal ionic liquid. Figure 7.8 shows the environmental interference functions of molten CuBr at 810 K, $Q\Delta i_{\mathrm{Cu}}(Q)$ for Cu and $Q\Delta i_{\mathrm{Br}}(Q)$ for Br, which are obtained from the measurements at incident energies of 8.680, 8.955, 13.170 and 13.445 keV. These four energies correspond to 300 and 25 eV below the Cu-K (8.980 keV) and Br-K (13.470 keV) absorption edges, respectively. The anomalous dispersion terms used for Cu are also listed in Table 7.1. The ordinary interference function, $Qi(Q)$, is estimated from the scattering profile at the single energy of 17.0 keV. On the other hand, Fig. 7.9 provides the three partial structure factors, $a_{\mathrm{ij}}(Q)$, for molten CuBr obtained by using a RMC simulation technique similar to that

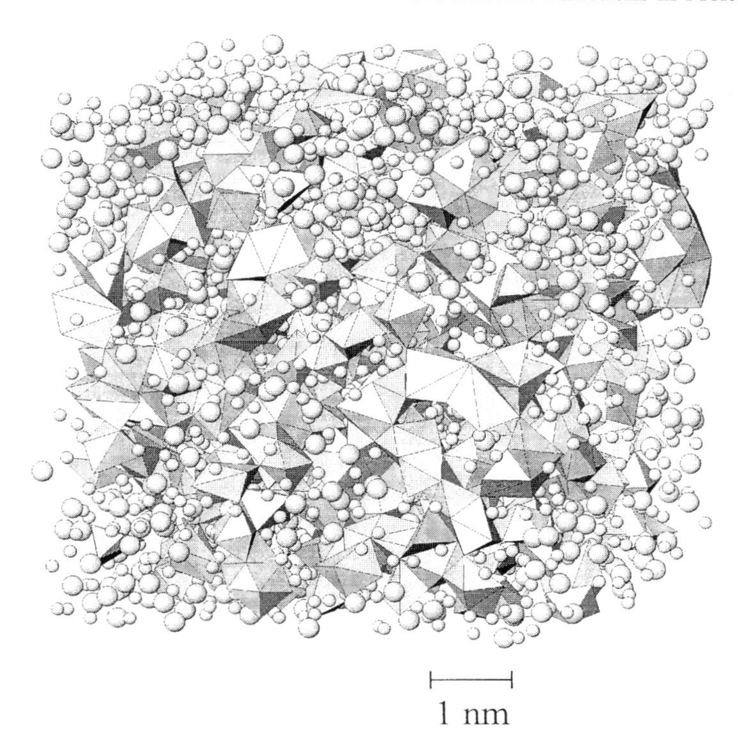


**Fig. 7.7.** A section of the atomic positions from a representative RMC-generated configuration for molten RbBr. *Larger atoms* are Br; *smaller atoms* are Rb. $RbBr_6$ units are selected and drawn as *octahedra*

used in the RbBr case. The vertical lines in this figure denote the uncertainty estimated from the experimental data by directly solving the simultaneous linear equation (7.4) [17]. The first trough of $a_{\mathrm{CuBr}}(Q)$ is located at a $Q$ value where the principal peak in $a_{\mathrm{BrBr}}(Q)$ is situated. In contrast, the partial structure factor $a_{\mathrm{CuCu}}(Q)$ for cation–cation pairs was found to be rather structureless, as was found in molten CuCl [2,11]. It is also worth mentioning that the present AXS results for molten CuBr surprisingly agree well with those obtained by the isotope substitution method for neutrons [15].

The partial pair distribution functions, $g_{ij}(r)$, of molten CuBr are given in Fig. 7.10. The closest Cu–Cu distance is significantly smaller than that for the Br–Br pair, indicating the like-ion penetration into the first unlike-ion coordination shell. This characteristic penetration may be interpreted as a decrease in the Coulombic repulsion interaction between copper ions arising from the reduced charge transfer between unlike ions. The peak positions and the coordination numbers estimated from these partial functions are given in Table 7.2, together with the values of the corresponding crystalline phase of

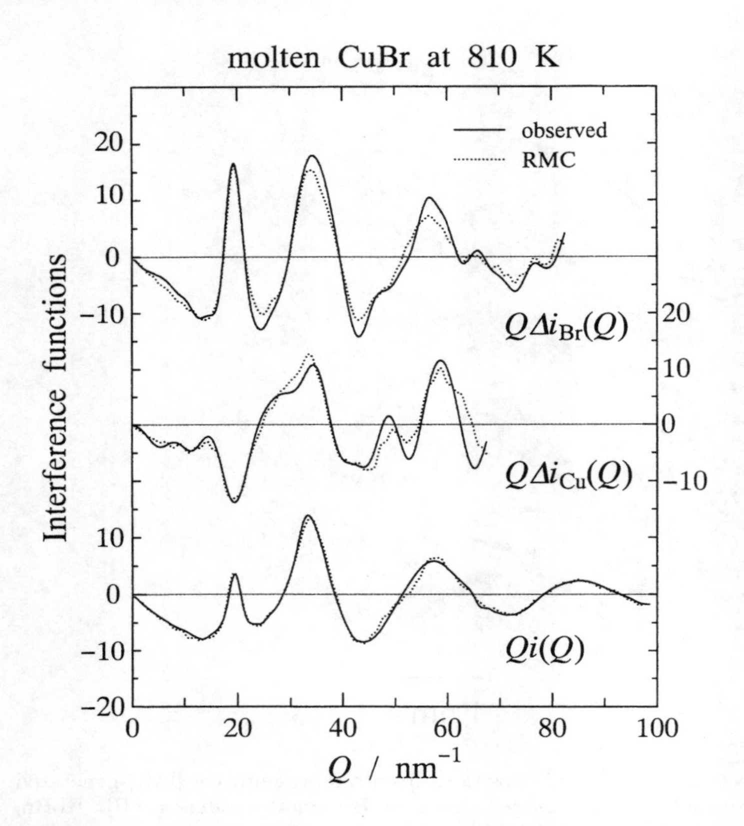


**Fig. 7.8.** The environmental interference functions $Q\Delta i_{\mathrm{Cu}}(Q)$ for Cu and $Q\Delta i_{\mathrm{Br}}(Q)$ for Rb in molten CuBr at 810 K, together with the ordinary interference function, $Qi(Q)$. *Solid lines* correspond to the experimental data. *Dotted lines* denote values calculated by the RMC simulation technique

CuBr of ZnS-type structure [28]. The coordination numbers for Br–Br pairs and Cu–Br pairs are found to be 11.4 at the distance of 0.398 nm and 3.1 at the distance of 0.245 nm, respectively. These values are more similar to those for a low-temperature crystalline phase, where the anions form a fcc lattice, than those for a high-temperature crystalline phase, possessing a bcc sublattice. In the molten state, the coordination number might be decreased by thermal agitation. However, the value of $r_{--}/r_{+-}$ for molten CuBr is 1.62, which is not far from that of the tetrahedral position (1.63). This is rather different from the octahedron value of 1.41. These AXS results imply that the structure of molten CuBr can be a disordered close packing of anions where the copper ions take a strongly disordered distribution by meandering through essentially tetrahedral holes.

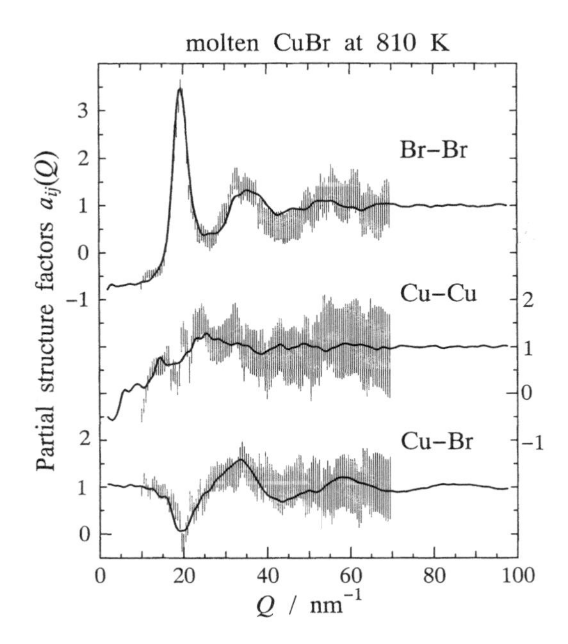


**Fig. 7.9.** Three partial structure factors of molten CuBr at 810 K. *Solid lines* correspond to the values calculated by the RMC simulation technique

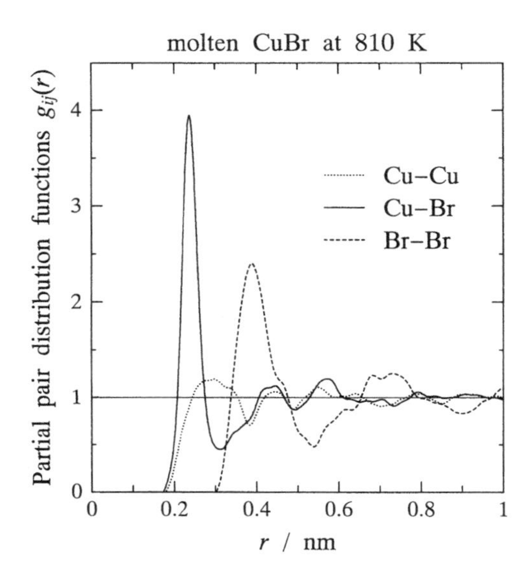


**Fig. 7.10.** Three partial pair distribution functions of molten CuBr at 810 K

Partial structural functions of molten AgBr [29] and CuI [30] are also estimated by coupling the AXS measurements with those of the conventional neutron diffraction without isotopes [31, 32]. It may also be worth mentioning that these numerical data are available from the public database (SCM-AXS) at http://www.tagen.tohoku.ac.jp. This includes the results of the liquid $Bi_{30}Ga_{70}$ alloy [33,34]. As easily seen in the normalized determinant of Fig. 7.4, the combination of AXS measurement and conventional neutron diffraction is an another way of obtaining the partial structure functions in binary systems. Considering many factors, the capability of the AXS method has been well recognized by obtaining the partial structure functions of a binary liquid, although the insufficient absolute accuracy of the experimental data still prevents us from obtaining mathematically exact solutions of the simultaneous linear equations directly from the AXS data alone. The RMC simulation technique appears to provide one useful way to reduce such an inconvenience.

## 7.2 Environmental Structures of Oxide and Metallic Glasses

The number of partial functions drastically increases with an increasing of number of components; three partial functions in a binary system, six partials in a ternary system, ten partials in quaternary system, etc.; it follows that there are $n(n+1)/2$ possible pairs in a system containing $n$ components. The separation of all partial functions is almost impossible from measured intensity data sets. This is particularly true in multi-component systems containing more than three elements.

In this subject, the concept of the environmental structure around a specific element is quite useful not only for non-crystalline systems but also for crystalline systems using the simple Fourier transformation, as already shown in Chap. 6. Of course, in this case, the basic definition of the partial structure is unchanged. Furthermore, the environmental structural analysis without complete separation of all partial functions has recently drawn much attention, because this data processing provides about an order of magnitude higher stability to the solutions than the direct AXS method [35, 19, 14].

It may be noted that the basic idea of the environmental structural analysis using the AXS measurement was suggested by Hosoya [36] and Shevchik [37], and its usefulness was clearly demonstrated with the results for amorphous $GeSe_2$ under the name of differential anomalous scattering (DAS) [35]. The basic conceptual approach of the environmental structural analysis for non-crystalline materials appears to be similar to the first-order difference scattering of neutrons with isotopes (see, for example, [38, 39, 40]). Many interesting results have recently been obtained, in parallel with the progress made using a synchrotron-radiation source. For this reason, some of the interesting results are given below.

### 7.2.1 Oxide Glasses

The binary $GeO_2$–$P_2O_5$ glass system is interesting, because of a change in environment around Ge with increasing $P_2O_5$ content. There have been some conflicting results [41, 42, 43, 44], and such disagreement may be due to the difficulty in determining the structure of Ge–O pairs in this binary system using the spectroscopic data alone. Then, the environmental structural analysis for Ge, namely whether Ge atoms prefer a tetrahedral site surrounded by four oxygen atoms or an octahedral site surrounded by six oxygen atoms, is strongly required. This prompts us to use the AXS method [45].

The ordinary RDFs computed from the interference functions of three $GeO_2$–$P_2O_5$ glass samples obtained by the conventional X-ray diffraction using Mo-K$\alpha$ radiation are illustrated in Fig. 7.11. Since this binary glass system consists of three components of Ge, P and O, the ordinary RDFs contain six partial functions of Ge–O, P–O, Ge–P, O–O, Ge–Ge and P–P pairs. From the ionic radius of the constituents, the first peak observed at about 0.18 nm may be allocated to the mixed correlation of the Ge–O and P–O pairs. However, the distances of tetrahedrally coordinated P–O (0.150 nm) and Ge–O (0.175 nm) pairs are close, so that the ordinary RDF analysis finds it difficult to separate the information for these two atomic pairs. The use of the AXS method reduces such difficulty by providing the environmental structure around a specific element, Ge in the present case, without complete separation of all partial functions.

The scattering intensity profiles of three samples were obtained from the AXS measurements at two energies near the Ge-K absorption edge in order to obtain the environmental interference functions for Ge, $Q\Delta i_{\mathrm{Ge}}(Q)$ and their RDFs. The results are shown in Figs. 7.12 and 7.13, respectively. The environmental functions contain only three partial functions of Ge–O, Ge–P and Ge–Ge pairs. Then, the first peak observed at about 0.18 nm in the environmental RDFs is attributed to the Ge–O pair correlation only. It may be worthy to note that the small peak around 0.24 nm is considered to be due to statistical fluctuations and enhancements of both Ge–O and Ge–Ge pairs. For convenience, the calculated values for Ge–O and Ge-Ge pairs are given by dotted lines in Fig. 7.13 using the case of $0.90GeO_2$–$0.10P_2O_5$ glass as an example. The coordination numbers of Ge–O pairs can be estimated from the corresponding peak area with the assumption that the peak shape is Gaussian. The results are summarized in Table 7.3 together with information for pure $GeO_2$ glass determined by the AXS method [46]. The coordination number and the distance of Ge–O pairs in the $GeO_2$–$P_2O_5$ glass increase with $P_2O_5$ content. Namely, the present AXS analysis clearly suggests that the variation of the coordinated oxygen atoms around Ge to be induced when adding the $P_2O_5$ component into the $GeO_2$ glass matrix. The AXS results are also found to be consistent with the conclusion obtained from infrared absorption [41] and Raman spectroscopy [42]. Therefore, we could interpret the increase in oxygen coordination number around Ge to be attributed to

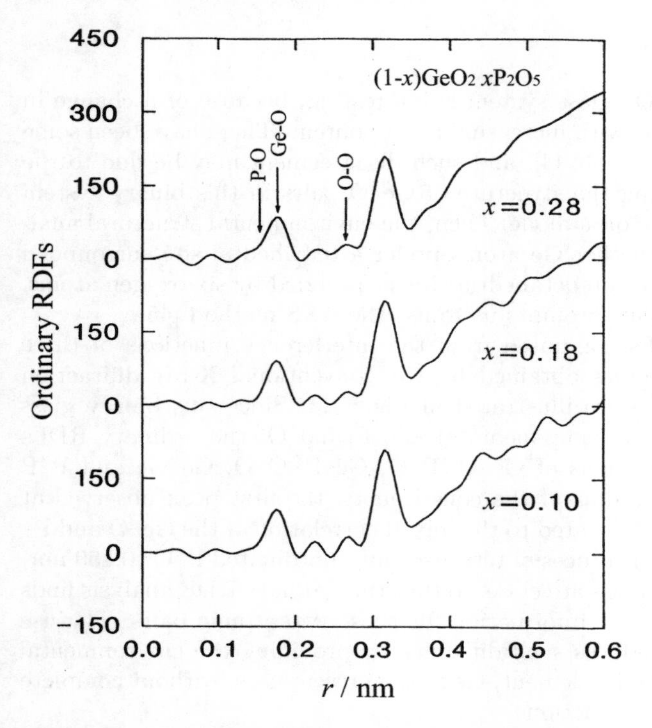


**Fig. 7.11.** The ordinary RDFs of $GeO_2$–$P_2O_5$ glasses

the mixed state of the tetrahedral and octahedral coordinations of Ge in the $GeO_2$–$P_2O_5$ glass.

Next, an example of a 1.4-μm-thick amorphous zinc ferrite film grown on a silica glass substrate produced by ion beam sputtering is described. In this case, we are relieved from the tedious correction procedure for the

**Table 7.3.** Comparison of the distance ($r$) and coordination number ($N$) for the Ge–O pairs of three $GeO_2$–$P_2O_5$ glasses determined by the AXS method

| Chemical composition | AXS measurement | | |
|---|---|---|---|
| | Ge–O | | Density |
| Mole fraction | $r$ (nm) | $N$ | ($Mg/m^3$) |
| $GeO_2$ | 0.175 | 4.1 | 3.64 |
| $0.90GeO_2$–$0.10P_2O_5$ | 0.175 | 3.9 | 3.61 |
| $0.82GeO_2$–$0.18P_2O_5$ | 0.177 | 5.1 | 3.59 |
| $0.72GeO_2$–$0.28P_2O_5$ | 0.182 | 5.3 | 3.55 |

Experimental uncertainty: $r$: ±0.002; $N$: ±0.4

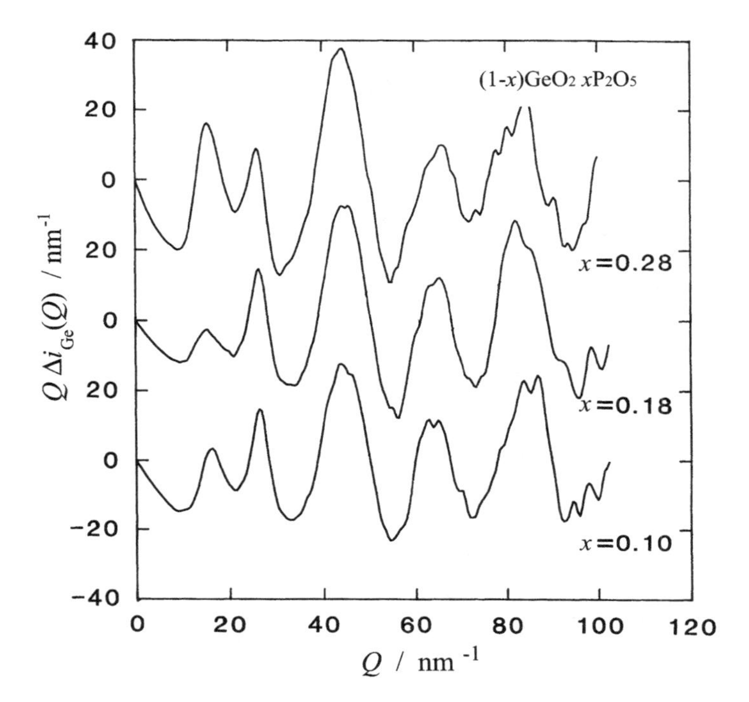


**Fig. 7.12.** Environmental interference function for Ge, $Q\Delta i_{\mathrm{Ge}}(Q)$, of $GeO_2$–$P_2O_5$ glasses

substrate material for the following reason: The contribution from a substrate material is automatically eliminated with systematic error cancellation when subtracting two data sets obtained with two energies close to the absorption edge of an element in a film (see Fig. 3.9).

The environmental interference function for Zn, $Q\Delta i_{\mathrm{Zn}}(Q)$, in the amorhous zinc ferrite film is obtained from the AXS measurements at energies of 9.635 and 9.360 keV below the Zn-K absorption edge (9.660 keV). Similarly, $Q\Delta i_{\mathrm{Fe}}(Q)$ for Fe is determined from the measurements at 6.811 and 7.086 keV below the Fe-K absorption edge (7.111 keV) [47]. The results are shown in Fig. 7.14, together with the ordinary interference function, $Qi(Q)$. On the other hand, Fig. 7.15 provides the RDFs computed from the interference function data.

Since six partial functions are involved in the ordinary RDF of Fig. 7.15, it is impossible to obtain information for Zn–O and Fe–O pairs with sufficient atomic sensitivity. On the other hand, the Fe–O pair contribution is not included in the environmental RDF for Zn or vice versa. Then, the first peaks of the environmental RDFs for Zn and Fe in Fig. 7.15 are ascribed to Zn–O and Fe–O pairs, respectively, when referring to the distance between nearest-

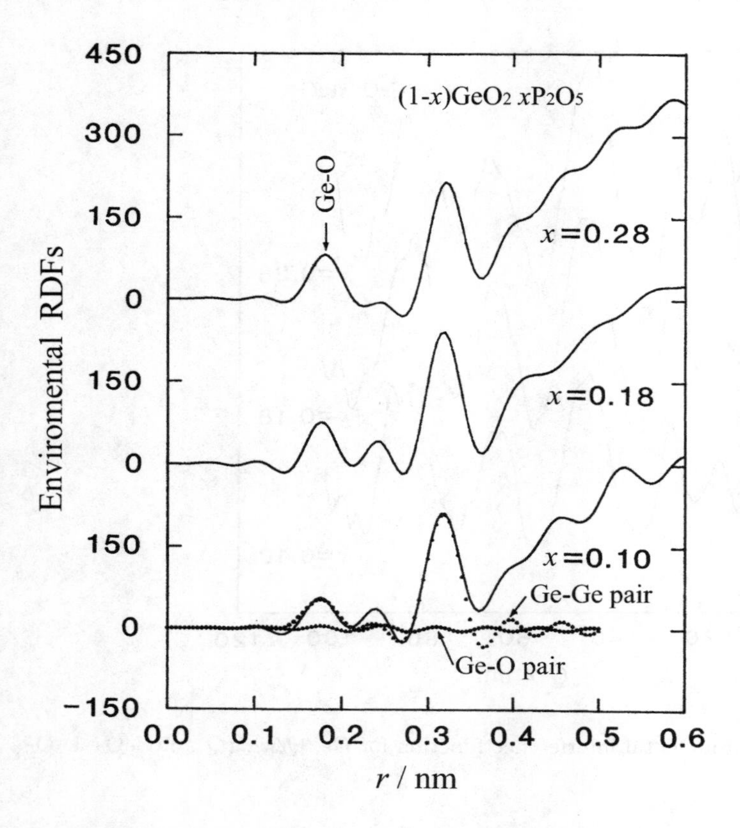


**Fig. 7.13.** Environmental RDFs for Ge of $GeO_2$–$P_2O_5$ glasses. *Dotted lines* for the $0.90GeO_2$–$0.10P_2O_5$ glass are an example of calculation using correlations of Ge–O and Ge–Ge pairs

neighboring oxygen and transition-metal elements in a zinc ferrite crystal with normal spinel structure. We can easily estimate from these results that the oxygen coordination numbers are $5.0 \pm 0.5$ for Zn and $5.8 \pm 0.5$ for Fe, respectively. This suggests that Zn is considered to occupy both tetrahedral (four coordination) and octahedral (six coordination) positions formed by oxygens in the amorphous film, whereas Fe is quite likely to occupy the octahedral site.

The structural parameters only for the nearest-neighboring pairs are not adequate for discussing structure–property relationships of non-crystalline materials. The disadvantage due to the limited wave vector range available is also appreciable, because of the relatively low-energy absorption edges of Zn and Fe in the present case. This prevents us from obtaining the RDF data with sufficient reliability due to finite termination in the Fourier transformation, and a careful interpretation for RDF is required. In this subject, the

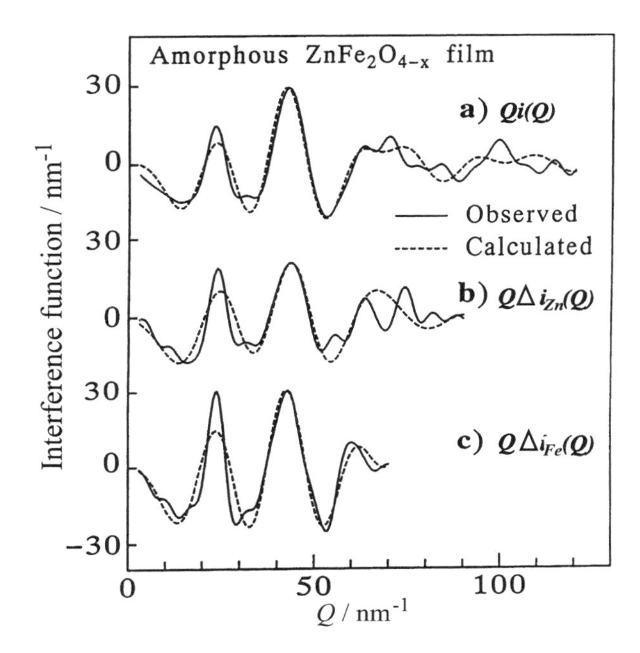


**Fig. 7.14.** Environmental interference functions for Zn and Fe together with the ordinary interference function of an amorphous zinc ferrite film

use of the least-squares refining technique first proposed by Levy, Dandord and Narten [48] may be one way to reduce such inconvenience, using the interference functions instead of RDFs. For convenience, the essential idea of the least-squares refining technique and its data processing are given below.

The essential idea of this approach [48,49,50] is based on the characteristic structural features of non-crystalline systems such as oxide glasses exemplified by the contrast between the narrow distribution of local ordering in the shorter-distance region and a complete loss of positional correlation in the longer-distance one. These concepts may be represented using the following equation with respect to the interference function:

$$\begin{aligned} Qi(Q) = &\sum_{j=1}\sum_{k} c_j \frac{f_j f_k}{\langle f \rangle^2} \frac{N_{jk}}{r_{jk}} \exp\Big(-b_{jk}Q^2\Big) \sin\Big(Q \cdot r_{jk}\Big) \\ &+ \sum_{\alpha}\sum_{\beta} \frac{c_\alpha c_\beta f_\alpha f_\beta}{\langle f \rangle^2} 4\pi\rho_\circ \\ &\times \exp(-b'_{\alpha\beta}Q^2) \frac{Qr'_{\alpha\beta}\cos(Q \cdot r'_{\alpha\beta}) - \sin(Q \cdot r'_{\alpha\beta})}{Q^2} . \end{aligned} \tag{7.10}$$

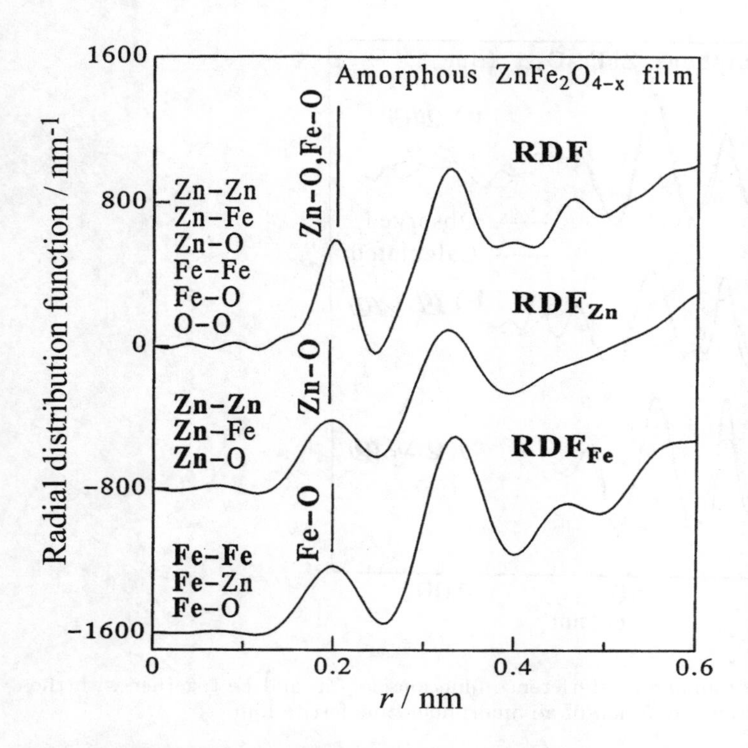


**Fig. 7.15.** Environmental RDFs for Zn and Fe together with the ordinary RDF of an amorphous zinc ferrite film

Here, $N_{jk}$ is the number of $k$ atoms around the origin atom $j$, $r_{jk}$ the average separation between $j$ and $k$, and $b_{jk}$ half of the mean-square variation in the Gaussian distribution. The first term in this equation represents the summation of Gaussian distributions within a certain distance of interest, $r'_{\alpha\beta}$, and the second term corresponds to the continuous distribution with an average number density of a system beyond the boundary region, $r'_{\alpha\beta}$ which need not be sharp [48, 49, 50]. In practice, the structural parameters are estimated by computing the interference function using the least-squares calculation of (7.10) so as to reproduce the experimental interference function. The environmental interference function can also be readily calculated by taking the difference of the calculated ordinary coherent scattering intensity data similarly estimated at the two energies and compared with the experimental AXS results. Thus, it is stressed here that the resolution of the structural parameters can be improved at a reasonable level by the least squares refining technique, when applying to not only $Qi(Q)$ but also $Q\Delta i(Q)$.

The structural parameters of near neighbors in the amorphous zinc ferrite film have been estimated by applying this procedure so as to reproduce three

**Table 7.4.** Coordination numbers and distances in an amorphous zinc ferrite film

| Pair | Spinel structure* | | AXS measurements | |
|---|---|---|---|---|
| | $r$ (nm) | $N$ | $r$ (nm) | $N$ |
| (tetrahedral site)–oxygen, (octahedral site)–oxygen | | | | |
| Zn–O | 0.203 | 5.6 | 0.202 ± 0.002 | 5.4 ± 0.5 |
| Fe–O | 0.201 | 5.2 | 0.198 ± 0.002 | 5.7 ± 0.4 |
| O–O | 0.298 | 12.0 | 0.311 ± 0.005 | 11.6 ± 1.8 |
| (octahedral site)–(octahedral site) | | | | |
| Zn–Zn | 0.298 | 1.8 | 0.295 ± 0.004 | 1.8 ± 0.3 |
| Zn–Fe | 0.298 | 2.9 | 0.310 ± 0.002 | 2.6 ± 0.2 |
| Fe–Fe | 0.298 | 2.3 | 0.295 ± 0.003 | 1.8 ± 0.2 |
| (tetrahedral site)–(octahedral site) | | | | |
| Zn–Zn | 0.349 | 2.1 | 0.351 ± 0.004 | 1.0 ± 0.3 |
| Zn–Fe | 0.349 | 5.3 | 0.338 ± 0.002 | 2.0 ± 0.2 |
| Fe–Fe | 0.349 | 5.7 | 0.341 ± 0.002 | 3.7 ± 0.2 |
| (tetrahedral site)–oxygen, (octahedral site)–oxygen | | | | |
| Zn–O | 0.360 | 8.9 | 0.371 ± 0.003 | 7.7 ± 0.7 |
| Fe–O | 0.357 | 9.6 | 0.353 ± 0.002 | 8.7 ± 0.4 |
| (tetrahedral site)–(tetrahedral site) | | | | |
| Zn–Zn | 0.364 | 0.2 | 0.369 ± 0.077 | 0.2 ± 0.3 |
| Zn–Fe | 0.364 | 0.7 | 0.360 ± 0.010 | 0.6 ± 0.2 |
| Fe–Fe | 0.364 | 1.2 | 0.370 ± 0.006 | 0.9 ± 0.2 |

* Inversion parameter $x = 0.78$

independent experimental data sets, as shown by dashed lines in Fig. 7.14. In this iterative approach, a convergence was searched for with several starting parameters evaluated from different site occupancy in the spinel structure by changing the ratio of Zn ions occupying the tetrahedral sites. The resultant numerical examples are listed in Table 7.4. It should be again stressed that these numerical values do not correspond to the mathematically unique solution. They are considered, at least, in the sense of the necessary condition to be best for explaining the three independent scattering intensity data, although they might be not the sufficient condition.

In principle, even two neighboring elements in the periodic table could be distinguished with the AXS method. However, practically in that case several important points must be kept in mind. They are given in the following example of an amorphous cobalt ferrite film showing spin glass behavior [51]. The local chemical environments around Fe and Co are necessary to understand this interesting magnetic property.

As shown in Fig. 7.16, the lower-energy side of the Co-K absorption edge just corresponds to the higher-energy side of the Fe-K absorption edge. Thus,

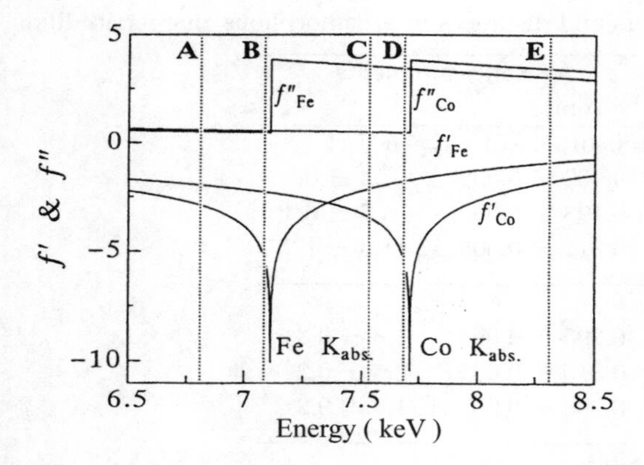


**Fig. 7.16.** Energy dependence of the anomalous dispersion terms (calculation) for Fe and Co in the energy region including the Fe- and Co-K absorption edges. The energies labeled A to E are used in the AXS measurements

the assumption used in the AXS method for the zinc ferrite case is not applicable for the cobalt ferrite film. More generalized forms for the AXS equations should be derived [52].

The energy variation of the real and imaginary parts of the anomalous dispersion term of the other elements as well as the element $A$ must be considered in such an AXS analysis. Then, the difference between the scattering intensities measured at two energies, $E_1$ and $E_2$, should be given in the following form, different from the simple form described in Chap. 2:

$$\Delta i(Q, E_1, E_2) = \frac{1}{W(Q, E_1, E_2)} \sum_{j=1}^{N} \sum_{k=1}^{N} c_j F_{jk}(Q, E_1, E_2) \times \int_0^\infty 4\pi r^2 [\rho_{jk}(r) - \rho_{0k}] \sin(Q \cdot r)/(Q \cdot r) \mathrm{d}r, \quad (7.11)$$

where

$$F_{jk}(Q, E_1, E_2) = \Re\Big[f_j(Q, E_1)\Big]\Re\Big[f_k(Q, E_1)\Big] + f_j''(E_1) f_k''(E_1) - \Re\Big[f_j(Q, E_1)\Big]\Re\Big[f_k(Q, E_2)\Big] - f_j''(E_2) f_k''(E_2), \quad (7.12)$$

$$W(Q, E_1, E_2) = \sum_{j=1}^{N} \sum_{k=1}^{N} c_j c_k F_{jk}(Q, E_1, E_2). \quad (7.13)$$

The term $\rho_{jk}(r)$ is the number density function of the $k$ component around $j$, and $\rho_{0j}$ the average number density for $j$. Fourier transformation of (7.11) gives the following environmental RDF:

$$4\pi r^2 \rho_\circ + \frac{2r}{\pi}\int_0^\infty Q\Delta i(Q, E_1, E_2)\sin(Q\cdot r)\mathrm{d}Q$$
$$= 4\pi r^2 \sum_{j=1}^{N}\sum_{k=1}^{N} \frac{c_j F_{jk}(Q, E_1, E_2)}{W(Q, E_1, E_2)}\rho_{jk}(r). \tag{7.14}$$

Figure 7.16 shows the anomalous dispersion terms of Fe and Co in the energy region including the Fe- and Co-K absorption edges. Three pairs of data sets labeled at A and B, C and D, and D and E in Fig. 7.16 were used to obtain the environmental RDFs. The energies of A to D are 6.811, 7.086, 7.534 and 7.684 keV, corresponding to 300 and 25 eV below the Fe-K absorption edge (7.111 keV) and 175 and 25 eV below the Co-K absorption edge (7.706 keV). The energy of E is 8.309 keV, corresponding to 600 eV above the Co-K absorption edge. It may be worth mentioning that two major problems have been recognized in this AXS measurement. The first problem is that the contributions from both Fe and Co overlap. For example, since the energy dependence of the anomalous dispersion terms is positive for $f'_{\mathrm{Fe}}$ and negative for $f'_{\mathrm{Co}}$ in the energy region between the two absorption edges, the intensity difference between C and D is substantially reduced. Second, because of the extended X-ray absorption fine structure (EXAFS) signal of Fe, the energy of C cannot be selected sufficiently close to the Fe-K absorption edge. This implies that a distinct change in $f'_{\mathrm{Co}}$ is hardly expected. In other words, we could use the relatively large variation in $f'_{\mathrm{Co}}$ and $f''_{\mathrm{Co}}$ in the AXS measurement at D and E, and then its intensity difference becomes larger at D and E than at C and D.

The three intensity profiles were obtained from the AXS measurements at A and B, C and D, and D and E. The resultant environmental RDFs are shown in Fig. 7.17 together with the ordinary RDF determined from the data with Mo-K$\alpha$ radiation [52]. As discussed above, the environmental structures for Fe and Co overlap in all three RDFs, although their contributions are varied in every RDF. Then, it is impossible to obtain information for the Fe–O and Co–O pairs directly from these environmental RDFs. Therefore, the coordination numbers of oxygen around Fe and Co are determined from the area under the first peak by solving four equations with relevant weighting factors for two unknown coordination numbers of the Fe–O and Co–O pairs. The results indicate that the oxygen coordination numbers are 5.2$\pm$0.6 for Fe and 6.3$\pm$0.4 for Co. The possible interpretation for the local ordering units in amorphous cobalt ferrite film may be suggested to be as follows: Nearly equal numbers of $FeO_4$ tetrahedra and $FeO_6$ octahedra exist as the fundamental local ordering units for the Fe environment, and the $CoO_6$ octahedral unit is preferred for the Co environment.

In conclusion, this results clearly demonstrate that structural information more than the average structure even in a system including neighboring elements in the periodic table can be determined by carefully applying the AXS method.

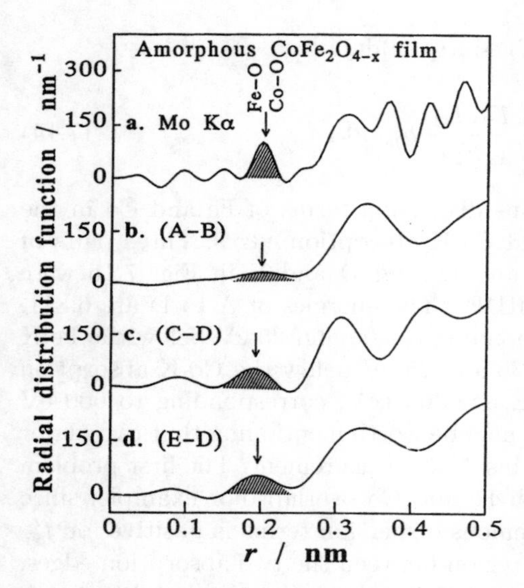


**Fig. 7.17.** Three environmental RDFs obtained from the AXS measurements at three pairs of energy sets: A and B, C and D and D and E, together with the ordinary RDF of amorphous cobalt ferrite film

### 7.2.2 Metallic Glasses

An extraordinary large number of metallic glasses (often called amorphous alloys or glassy metals) have been produced by several methods such as rapid quenching from melt and vapor deposition. This growing field was strongly stimulated by the report of Duwez and his colleagues in 1960 [53] with respect to a $Au_{70}Si_{30}$ glass sample prepared from the melt using the piston and anvil technique. The production by Masumoto and Maddin in 1971 [54] of a $Pd_{80}Si_{20}$ glass sample with ribbon form and the development of a single-roller quenching technique by Chen and his colleague [55] have also led to many new advances in this field. Thus, the technological potentials of metallic glasses has now been recognized for application in soft magnetic elements, electronic devices and excellent high-strength wires with good corrosion resistance. The distinguishing feature of these relatively new materials can be characterized by the lack of crystal-like atomic periodicity.

Current interest in this field arises mainly from the discovery of bulk amorphous alloys by Inoue and his colleagues [56, 57, 58]. This new finding made production of a metallic glass ingot as large as 72 mm in diameter and 75 mm in length possible [59]. For this reason, the following provides some structural results obtained from the AXS measurements for particular metallic glasses with a wide supercooled liquid region, $\Delta T_x$, which is defined by the difference between the crystallization temperature, $T_x$ and the glass transition temperature $T_g$ [60, 56]. It may be stressed here that good glass

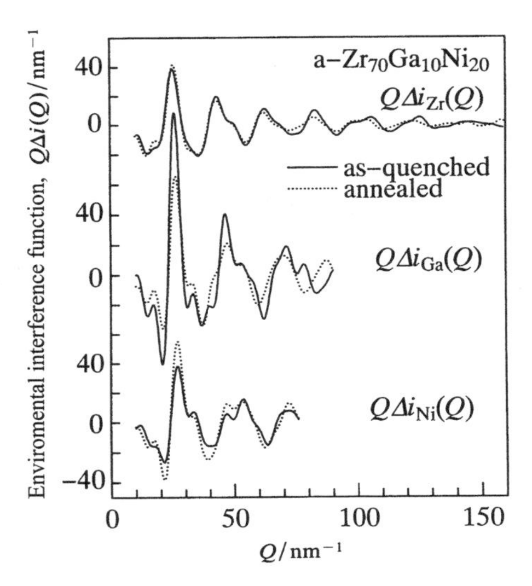


**Fig. 7.18.** Environmental interference functions for Zr, Ga and Ni of the $Zr_{70}Ga_{10}Ni_{20}$ metallic glass

formability is obtained, when alloys show a large value for $\Delta T_x$, as well as for the ratio of $T_g/T_m$, where $T_m$ is the melting temperature. In addition, metallic glasses with good glass-forming ability always contain more than three elements with large, medium and small atomic sizes.

Figure 7.18 provides the environmental interference functions for all three glass components of the $Zr_{70}Ga_{10}Ni_{20}$ metallic glass ($\Delta T_x = 50\,K$, $T_g = 674\,K$ and $T_x = 724\,K$). Solid lines denote the as-quenched sample, and dotted lines correspond to the sample annealed for 120 s at 700 K in the supercooled liquid region. These results were obtained from the AXS measurements at two energies, which were 30 and 300 eV below the Zr (17.998 keV), Ga (10.367 keV) and Ni (8.333 keV) energies, respectively [61]. The environmental RDFs are shown in Fig. 7.19 together with the ordinary RDF computed from the intensity data measured at an energy of 300 eV below the Zr-K absorption edge. The ordinary RDF is a weighted sum of the six partial functions for Zr–Zr, Zr–Ga, Zr–Ni, Ga–Ga, Ga–Ni and Ni–Ni pairs, whereas the environmental RDF for Zr, Ga and Ni contains a weighted sum of the three partial functions for pairs around Zr, Ga and Ni, respectively. Therefore, by considering the Goldschmidt radii of each element (Zr: 0.160 nm, Ga: 0.141 nm and Ni: 0.124 nm), the first peaks in the environmental RDFs are ascribed to the pairs indicated by arrows in Fig. 7.19. Although the absolute values of the ordinary and environmental RDFs provide different physical meanings, the following points may be suggested: The first peak in the environmental RDF for Zr has a shoulder at the lower $r$-side, which resembles the peak profile in the ordinary RDF. This is explained by the fact that the contribution

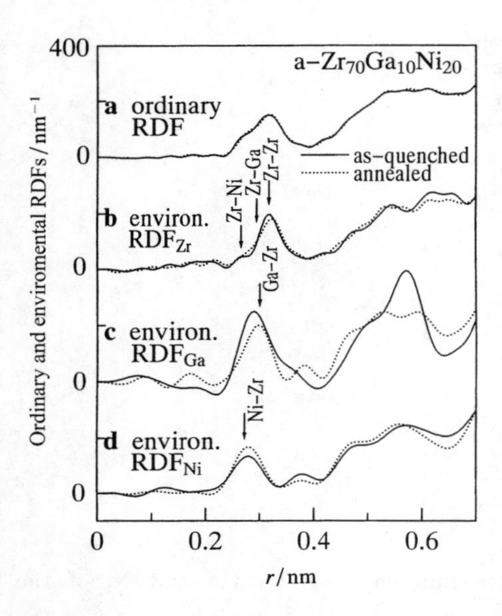


**Fig. 7.19.** Environmental RDFs for Zr, Ga and Ni of the $Zr_{70}Ga_{10}Ni_{20}$ metallic glass together with the ordinary RDF

of the pairs including Zr is much larger than the others, because of larger scattering ability and concentration of Zr in this alloy system. On the other hand, it may safely be concluded that the first peaks in the environmental RDF for Ga and Ni are attributed to Ga–Zr and Ni–Zr pairs, respectively. Therefore, the distances and coordination numbers of Zr around Ga and Ni in the nearest-neighbor region are estimated from these peaks. Using these values for Zr–Ga and Zr–Ni pairs, the coordination number for Zr–Zr pairs could be estimated from the first peak of the environmental RDF for Zr. The results for the as-quenched and annealed samples are summarized in Table 7.5.

It is shown in this table that the coordination numbers of all three components around Zr change markedly, and only the distance of Zr–Ga pairs apparently increases by annealing in the supercooled liquid region. Further-

**Table 7.5.** Coordination numbers and distances in the $Zr_{70}Ga_{10}Ni_{20}$ metallic glass as-quenched and annealed for 120 s at 700 K

| Samples | Zr–Zr | | Zr–Ga | | Zr–Ni | |
|---|---|---|---|---|---|---|
| | $r$ (nm) | $N$ | $r$ (nm) | $N$ | $r$ (nm) | $N$ |
| As-quenched | 0.322 | 6.6 | 0.288 | 1.3 | 0.276 | 1.3 |
| Annealed | 0.325 | 7.0 | 0.292 | 0.9 | 0.277 | 1.9 |

Experimental uncertainty: $r$: ±0.002; $N$: ±0.3

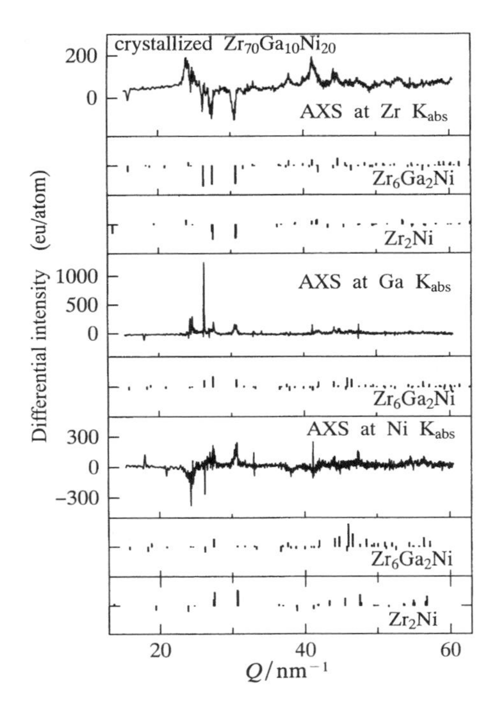


**Fig. 7.20.** The AXS intensity profiles for the crystallized $Zr_{70}Ga_{10}Ni_{20}$ alloy at the K absorption edges of Zr, Ga and Ni. A comparison is made with those for the crystalline phases of $Zr_2Ni$ and $Zr_6Ga_2Ni$

more, the precipitated phases have been determined to be a mixture of $Zr_2Ni$ and $Zr_6Ga_2Ni$ by applying the AXS method to all three glass components, as shown in Fig. 7.20, in the crystallized sample prepared by annealing for a long time in the supercooled liquid region. The average distances and coordination numbers of two crystalline phases of $Zr_2Ni$ and $Zr_6Ga_2Ni$ are listed in Table 7.6. Taking account of the precipitated crystalline phases [62, 63] and the variation in the Zr environment, including a relatively large increase detected in the distance of Zr–Ga pairs, it is concluded that the redistribution of Ga in the Zr environment is essentially required for crystallization in the $Zr_{70}Ga_{10}Ni_{20}$ metallic glass.

Similar AXS works were carried out for metallic glasses with a wide supercooled liquid region. This includes the results for particular metallic glasses of $Zr_{70}Al_{15}Ni_{25}$ [64], $La_{55}Al_{25}Ni_{20}$ [65] and $Zr_{33}Y_{27}Al_{15}Ni_{25}$ [66]. For example, in the $Zr_{70}Al_{15}Ni_{25}$ metallic glass, the arrangement of Al around Zr differs markedly from those in the crystallized sample. This again implies that a large change in the Zr environment is indispensable for crystallization, which provides good thermal stability of this particular metallic glass.

**Table 7.6.** Coordination numbers and distances in the near-neighbor region for the precipitated phases of $Zr_6Ga_2Ni$ and $Zr_2Ni$ in the $Zr_{70}Ga_{10}Ni_{20}$ alloy crystallized by annealing for 180 ks at 700 K in the supercooled liquid region

| Pairs | $Zr_6Ga_2Ni$ | | $Zr_2Ni$ | |
|---|---|---|---|---|
| | $r$ (nm) | $N$ | $r$ (nm) | $N$ |
| Zr–Zr | 0.327 | 9 | 0.331 | 11 |
| Zr–Ga | 0.299 | 3 | – | – |
| Ga–Zr | 0.299 | 9 | – | – |
| Zr–Ni | 0.275 | 1.5 | 0.276 | 4 |
| Ni–Zr | 0.275 | 9 | 0.276 | 8 |
| Ni–Ni | | | 0.263 | 2 |

Although the origin of a wide supercooled liquid region and good glass formability in these multi-component metallic glasses cannot be identified with certainty yet, the available AXS results suggest the following points:

(1) The local ordering structure in these metallic glasses markedly differs from those found in the precipitated crystalline phases. In contrast, the local ordering structure in metallic glasses without a wide supercooled liquid region is found to resemble those of the corresponding equilibrium crystalline phases.
(2) These experimental observations confirm the presence of a close relation between the local ordering structure and the thermal stability in these particular metallic glasses.
(3) These thermally stable metallic glasses are decomposed into more than two types of crystalline phases. This also certainly makes redistribution of the glass components extremely difficult and retards crystallization.

## 7.3 Solutions

Various iron oxides, oxy-hydroxides and hydroxides are widely distributed in soils, minerals, river, sea water and play an important role in our environment, biological actions and industrial activities. The formation and transformation of hydrous oxides in aqueous solutions are reviewed, for example, by Blesa and Matijevic [67]. However, little is known about the atomic-scale structure of ferric hydroxides in solutions, because an experimental difficulty lies in the coexistence of a large amount of water molecules. Similarly, the structure around a metallic cation in aqueous solutions is difficult to determine with both conventional X-ray and neutron diffraction.

For example, in a metal-halide solution, such as the $ZnCl_2$–$H_2O$ system, the scattering intensity data obtained by the conventional method include a sum of ten partial structure factors, which also contains a large contribution from the partial structure factor of O–O pairs. In this subject, valuable simplification can be made by tuning the energy of X-rays in the close vicinity of the absorption edge for a specific element, Zn in the present case. The environmental structure around Zn determined using the AXS method consists of only four partial structure factors without a large contribution from water molecules. This enables us to evaluate a precise atomic structure around a metallic cation in solutions. Such an idea has been presented [68], in a manner similar to the first-order difference scattering of neutrons with isotopes (see, for example, [38, 39]).

The use of the AXS method also has another merit in the experiments for solutions (see also Fig. 3.9). For diffraction experiments, a solution sample is usually contained in a cell with windows transparent to X-rays. For quantitative analyses, the intensity from a solution should be accurately subtracted for the scattering intensity from the window materials. Such data reduction is done using the intensity only from the window materials and correction for absorption due to solution. In the AXS measurements at two different energies, each scattering intensity profile contains the contribution from the window materials as well as from the solution. By taking the difference between the two profiles, however, the contribution from the window materials as well as that from the non-zinc pairs is automatically eliminated. In this way, we are released from the tedious correction procedure for the window materials.

In this section, two selected results for aqueous solutions by the AXS method are given, in order to demonstrate the potential of the AXS method. This includes a simple hydration structure in $ZnCl_2$ aqueous solutions [69] and the more complicated structural behavior of poly-molybdate ions in acid Mo–Ni solutions [70]. Nevertheless, it should be stressed here that the first AXS application to solutions using the synchrotron radiation source has been made by Ludwig et al. [71] with respect to aqueous solutions of three transition metal bromides ($ZnBr_2$, $CuBr_2$, and $NiBr_2$), under the name of differential anomalous scattering (DAS). They reported, for example, the formation of tetrahedral complexes around $Zn^{2+}$ in an aqueous $ZrBr_2$ solution and an octahedral coordination shell in an aqueous $CuBr_2$ solution. In contrast, there are only a few Ni–Br nearest neighbors and cations and anions share hydrating water molecules in $NiBr_2$ solution.

### 7.3.1 Hydration Structure of $ZnCl_2$ Aqueous Solutions

Figure 7.21 shows the intensity profiles of 0.98 and 2.85 mol $l^{-1}$ $ZnCl_2$ aqueous solutions measured at energies of 9.361 and 9.636 keV, just below the Zn-K absorption edge (9.660keV) [69]. The energy differential profile between the two intensities, $\Delta I_{Zn}$, in each solution is also shown in this figure. By analyzing the AXS results of $\Delta I_{Zn}$, the environmental RDF for Zn in solutions can be estimated using the common Fourier transformation, and the results

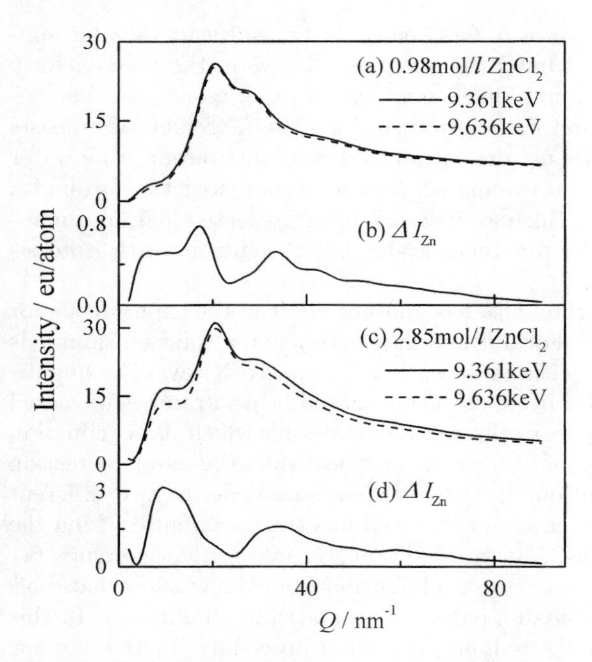


**Fig. 7.21.** Intensity profiles of the **a** 0.98 mol $l^{-1}$ and **b** 2.85 mol $l^{-1}$ $ZnCl_2$ aqueous solutions obtained from the AXS measurements at the Zn-K absorption edge. **c,d** Respective intensity differences

are given in Fig. 7.22. The broken lines in this figure correspond to the ordinary RDFs, which reflect mainly the contribution from water molecules as well as the average of ten partials. It may be again stressed here that a direct comparison between the two RDF curves is only possible for the values on the abscissa, because different weighting factors are used on the vertical scale.

The peak at about 0.3 nm in the ordinary RDF, which is ascribed to O–O pairs of water molecules [72] and $Cl^-$–O pairs of a water molecule hydrated around a chloride ion [73], disappears in the environmental RDFs. Referring to the distance between a zinc ion and oxygen of hydrated water molecules reported in the literature, the hydration number and distance of zinc ions can be estimated from the first peak in the environmental RDFs. Then, the hydration numbers around zinc are 5.7 ± 0.7 at 0.210 ± 0.002 nm in the 0.98 mol $l^{-1}$ $ZnCl_2$ solution and 6.2 ± 0.2 at 0.215 ± 0.002 nm in the 2.85 mol $l^{-1}$ $ZnCl_2$ solution, respectively. The formation of tetrahedral complexes around a zinc ion is suggested in some aqueous solutions containing zinc ions [71, 74]. However, the AXS results for aqueous $ZnCl_2$ solutions agree well with those obtained by the conventional X-ray diffraction method [75, 76, 77, 78] and the EXAFS case [79]. Namely, the hydration number of a zinc ion is about six and the hydration distance is about 0.210 nm. The

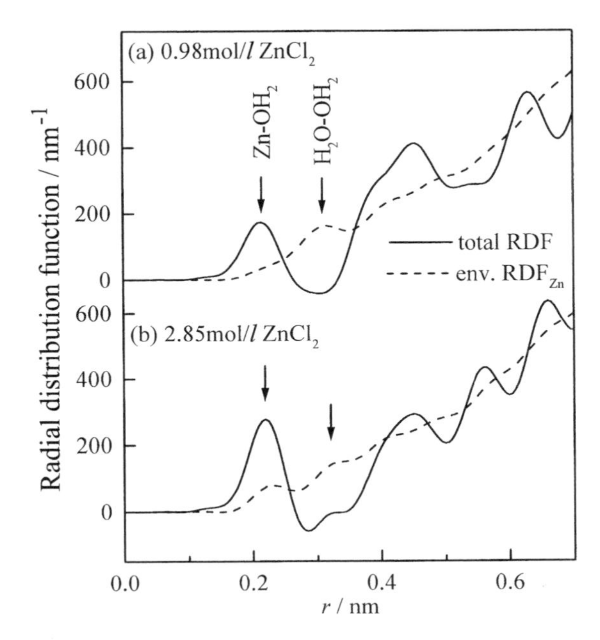


**Fig. 7.22.** Environmental RDFs for Zn (*solid line*) and ordinary RDFs (*broken line*) in the **a** 0.98 mol $l^{-1}$ and **b** 2.85 mol $l^{-1}$ $ZnCl_2$ aqueous solutions

hydration number and distance in the 2.85 mol $l^{-1}$ $ZnCl_2$ solution coincide with the values for the 0.98 mol $l^{-1}$ $ZnCl_2$ case within the experimental uncertainties. Then, the hydration structure around a zinc ion is considered to be insensitive to the solute concentration by finding no significant change even in the 2.85 mol $l^{-1}$ $ZnCl_2$ solution. Nevertheless, the possible interpretation of the data by postulating the formation of some kind of complexes containing $Cl^-$ around $Zn^{2+}$ is not excluded.

The hydration structure of a heavier element, such as Er, has also been determined by the AXS method using the $L_{III}$ absorption edge instead of the K absorption edge [80]. Coherent scattering intensities observed at 8.339 and 8.064 keV below the Er-$L_{III}$ absorption edge (8.3575 keV) in 0.5 and 1.0 mol $l^{-1}$ aqueous $ErCl_3$ solutions are given in Fig. 7.23. It is noted that average intensity differences near the first peak are, for example, about 10 and 15% for the in 0.5 and 1.0 mol $l^{-1}$ aqueous $ErCl_3$ solutions. Since the magnitude of anomalous dispersion term, $f'$, is about two or three times larger at the $L_{III}$ absorption edge than at the K absorption edge, a larger intensity difference can be achieved when using the $L_{III}$ absorption edge. This sometimes enables us to obtain the AXS data even from an element of very minor concentration. This includes the results for the concentration of 0.5 mol $l^{-1}$ $\sim$ 0.3 at. % Er in aqueous solution.

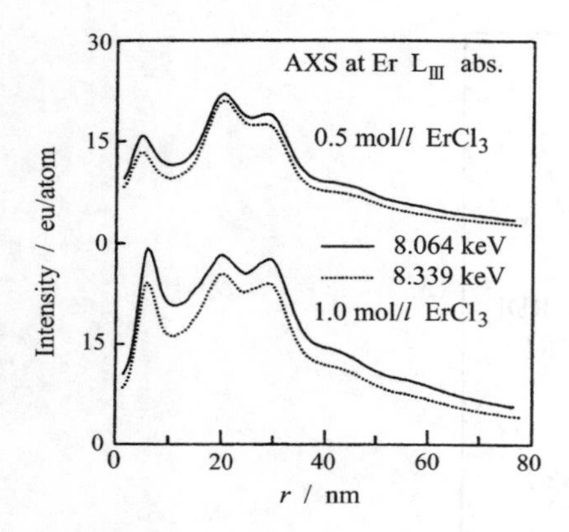


**Fig. 7.23.** Intensity profiles of the **a** 0.5 mol $l^{-1}$ and **b** 1.0 mol $l^{-1}$ $ErCl_3$ aqueous solutions obtained from AXS measurements at the Er-$L_{III}$ absorption edge

The environmental RDFs are shown in Fig. 7.24 together with the ordinary RDFs denoted by dotted lines. In the ordinary RDFs, three peaks at about 0.24, 0.30 and 0.35 nm indicated by arrows are observed. These first and second peaks are attributed to $Er^{3+}$–O and O–O pairs. Then, the hydration numbers estimated from the area under the first peak of the environmental RDFs for erbium are $8.2 \pm 0.6$ at $0.240 \pm 0.002$ nm in the 0.5 mol $l^{-1}$ $ErCl_3$ solution and $8.3 \pm 0.3$ at $0.238 \pm 0.002$ nm in the 1.0 mol $l^{-1}$ $ErCl_3$ solution. On the other hand, the hydration around $Cl^-$ in the 2.85 mol $l^{-1}$ aqueous $NdCl_2$ heavy water solution was determined using the isotope substitution technique of neutrons [81]. The distance between $Cl^-$ and O of heavy water ($D_2O$) molecules is estimated to be $0.345 \pm 0.004$ nm. Based on such information, the third peak observed at 0.35 nm in the ordinary RDF of Fig. 7.24 may result mainly from the Cl–O pairs, which is consistent with the fact that this peak almost disappears by taking the difference. This result clearly suggests not only successful application of the AXS method to a study of the hydration structure in aqueous solutions, but also the advantage of using the $L_{III}$ absorption edge for the AXS measurements.

### 7.3.2 Poly-molybdate Complexes in Acid Mo–Ni Solutions

Electrodeposited molybdenum or tungsten amorphous alloys, such as Mo–Ni [82] and W–Co [83], provide superior corrosion and wear resistance. Their catalytic activity in the industrial production of hydrogen from alkaline-water electrolytic cells is also notable [84]. In addition to this practical relevance, the electrodeposition of molybdenum or tungsten alloys is of considerable interest from a scientific point of view. Pure molybdenum or tungsten is not electrodeposited from an aqueous solution, whereas it is electrically co-deposited only from an aqueous solution containing an iron-group metal such

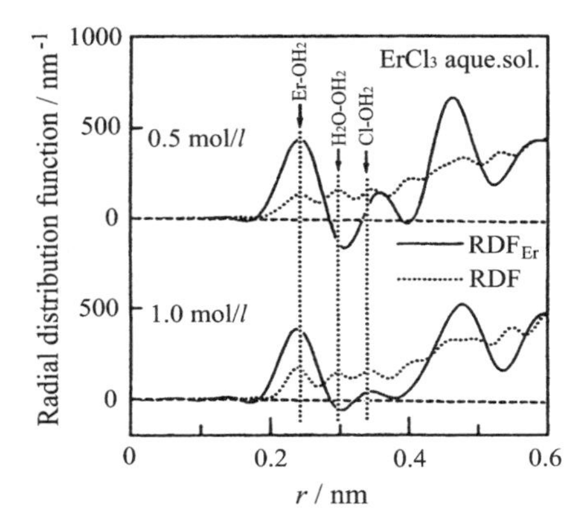


**Fig. 7.24.** Environmental RDFs for Er (*solid line*) and ordinary RDFs (*dotted line*) in the **a** 0.5 mol $l^{-1}$ and **b** 1.0 mol $l^{-1}$ $ErCl_3$ aqueous solutions.

as Fe and Ni. Brenner [85] explained the electrodeposition of these alloys by introducing the concept of induced co-deposition.

Several theoretical assumptions [86, 87, 88, 89, 90] have been made to explain the mechanism of induced co-deposition. The electroplating conditions [83,89,90,91], the mechanical and chemical properties [92,93,94] and the atomic or micrographic structures [84,95] of the co-deposited molybdenum or tungsten alloy have been intensively studied. However, the mechanism of induced co-deposition is not well understood yet. In order to obtain a new scope for the induced co-deposition process, the environmental structure around Mo or Ni in Mo–Ni aqueous solutions has been determined by using the AXS and extended X-ray absorption fine structure (EXAFS) methods [70]. In this work, the following points are well organized: Structural information obtained from the EXAFS result is sensitive particularly to the first nearest-neighbor pairs. On the other hand, the AXS method provides structural information for atomic pairs even at a greater distance. For this reason, the structural parameters for adjacent Mo–O and Ni–O pairs at the first nearest-neighbor distance are determined by the EXAFS method. Using such EXAFS data for Mo–O and Ni–O pairs, the structural parameters for neighboring Mo–Mo and Mo–Ni pairs at a greater distance are then determined by the least-squares refining method coupled with the differential intensity function obtained by the AXS method.

Figure 7.25 shows the Fourier transform of weighted EXAFS spectra for Mo ions in 1.0 mol $l^{-1}$ $Na_2MoO_4$, 0.5 mol $l^{-1}$ $Na_2MoO_4$+0.5 mol $l^{-1}$ $NiSO_4$, and 0.5 mol $l^{-1}$ $Na_2MoO_4$ + 0.5 mol $l^{-1}$ $NiSO_4$ + 1.0 mol $l^{-1}$ sodium citrate dihydrate ($Na_3C_6H_5O_7 \cdot 2H_2O$) aqueous solutions. No significant difference is detected in the peaks at about 0.12 and 0.18 nm in all solutions. Then, these Mo radial structure functions (RSFs) suggest that the first neighboring

**Table 7.7.** Coordination numbers and distances in the first neighboring shell determined from the Mo EXAFS spectra in 1.0 mol $l^{-1}$ $Na_2MoO_4$, 0.5 mol $l^{-1}$ $Na_2MoO_4$ + 0.5 mol $l^{-1}$ $NiSO_4$, and 0.5 mol $l^{-1}$ $Na_2MoO_4$ + 0.5 mol $l^{-1}$ $NiSO_4$ + 1.0 mol $l^{-1}$ $Na_3$ cit. aqueous solutions, and $MoO_3$ crystal [96]. Experimental errors: $N$: ±0.2; $r$: ±0.002 nm, respectively. $Na_3$ cit: $Na_3C_6H_5O_7 \cdot 2H_2O$

| Aqueous solution | Mo–O(1) | | Mo–O(2) | | Mo–O(3) | |
|---|---|---|---|---|---|---|
| | $r$ (nm) | $N$ | $r$ (nm) | $N$ | $r$ (nm) | $N$ |
| 1.0 mol $l^{-1}$ $Na_2MoO_4$ | 0.173 | 2.1 | 0.193 | 2.0 | 0.224 | 2.0 |
| 0.5 mol $l^{-1}$ $Na_2MoO_4$ + 0.5 mol $l^{-1}$ $NiSO_4$ | 0.172 | 2.1 | 0.192 | 2.2 | 0.223 | 1.9 |
| 0.5 mol $l^{-1}$ $Na_2MoO_4$ + 0.5 mol $l^{-1}$ $NiSO_4$ + 1.0 mol $l^{-1}$ $Na_3$ cit. | 0.172 | 2.0 | 0.194 | 2.1 | 0.228 | 2.0 |
| $MoO_3$ crystal | 0.170 | 2.0 | 0.194 | 2.0 | 0.229 | 2.0 |

shells around Mo ions are not significantly influenced by other ionic species. Coordination numbers and atomic distances for the first shell around Mo ions were determined by the Fourier filtering technique for EXAFS. The region denoted by the dashed lines in Fig. 7.25 corresponds to the width of a window function used for filtering the first neighboring shell. The Fourier-filtered Mo EXAFS spectra, $k^3\chi(k)$, are provided by the solid lines in Fig. 7.26, together with the calculated values, denoted by the dashed lines in this figure, from the structural parameters in Table 7.7. It is worth mentioning that a model with three Mo–O pairs with different distances is required to fully fit the Mo EXAFS spectra. This is supported by the EXAFS spectrum of crystalline $MoO_3$ as shown in Fig. 7.26d. According to the crystalline molybdenum trioxide $MoO_3$ data [96], the structural unit is a distorted octahedron consisting of a molybdenum atom surrounded by six oxygen atoms. These six oxygen atoms are grouped in twos with three different Mo–O distances of 0.167–0.173 nm, 0.195 nm and 0.225–0.233 nm. These values are consistent with the EXAFS results in crystalline $MoO_3$ in Table 7.7. Therefore, the Mo EXAFS spectra of Mo–Ni solutions in Fig. 7.26 have been analyzed with the model of three Mo–O distances, as is seen in Table 7.7. This clearly supports the fact that the Mo ions form the distorted $MoO_6$ octahedra even in the solutions.

Figure 7.27 shows the environmental RDFs around Mo ions in three aqueous solutions determined from the AXS data. From the structural data of Mo–O pairs in the near-neighbor region summarized in Table 7.7, we may conclude that two peaks at about 0.18 and 0.23 nm in the RDFs in Fig. 7.27 are attributed to the overlapped Mo–O(1) and Mo–O(2) pairs and the Mo–O(3) pairs, respectively. As clearly seen in the EXAFS results of Fig. 7.25, any distinct peak is not detected in the region of 0.33 nm, in which an apparent peak of Mo–Mo pairs is observed in the $MoO_3$ crystal. Quantitative analysis of Mo–Mo(Ni) pairs in solutions is very difficult from the EXAFS data alone. On the other hand, the distinct peak of Mo–Mo(Ni) pairs is clearly detected at about 0.33 nm in the environmental RDFs for Mo. Thus, the structural parameters for Mo–Mo(Ni) pairs in these solutions can be accurately deter-

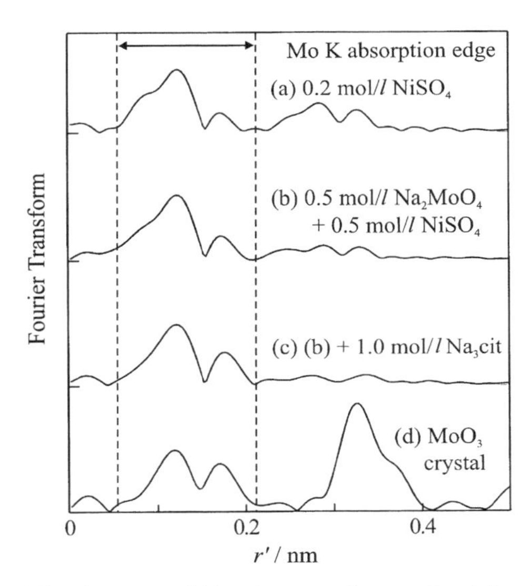


**Fig. 7.25.** **a–d** Fourier transforms of weighted EXAFS spectra in aqueous solutions. **a** 1.0 mol $l^{-1}$ $Na_2MoO_4$, **b** 0.5 mol $l^{-1}$ $Na_2MoO_4$ + 0.5 mol $l^{-1}$ $NiSO_4$, **c** 0.5 mol $l^{-1}$ $Na_2MoO_4$ + 0.5 mol $l^{-1}$ $NiSO_4$ + 1.0 mol $l^{-1}$ sodium citrate dihydrate and **d** $MoO_3$ crystal

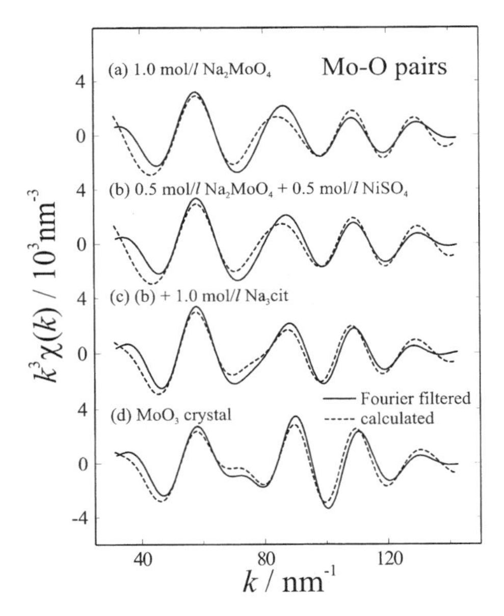


**Fig. 7.26.** **a–d** Fourier-filtered Mo EXAFS spectra in aqueous solutions. **a** 1.0 mol $l^{-1}$ $Na_2MoO_4$, **b** 0.5 mol $l^{-1}$ $Na_2MoO_4$ + 0.5 mol $l^{-1}$ $NiSO_4$, **c** 0.5 mol $l^{-1}$ $Na_2MoO_4$ + 0.5 mol $l^{-1}$ $NiSO_4$ + 1.0 mol $l^{-1}$ sodium citrate dihydrate and **d** $MoO_3$ crystal. The *dashed lines* denote the calculated values using the data listed in Table 7.7

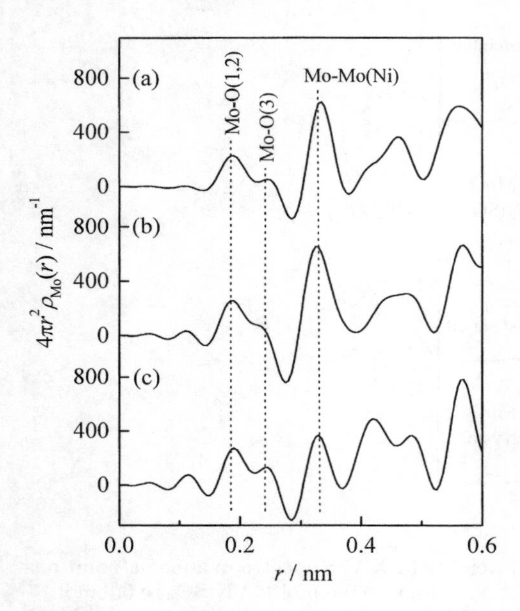


**Fig. 7.27.** **a–c** Environmental RDFs for Mo in aqueous solutions. **a** 1.0 mol $l^{-1}$ $Na_2MoO_4$, **b** 0.5 mol $l^{-1}$ $Na_2MoO_4$ + 0.5 mol $l^{-1}$ $NiSO_4$, **c** 0.5 mol $l^{-1}$ $Na_2MoO_4$ + 0.5 mol $l^{-1}$ $NiSO_4$ + 1.0 mol $l^{-1}$ sodium citrate dihydrate

mined from the AXS data using the Mo-K absorption edge. The coordination numbers and atomic distances at about 0.33 nm were determined from the structural parameters for the Mo–O pairs in Table 7.7 so as to reproduce the differential intensity function for Mo, $Q\Delta i_{\mathrm{Mo}}(Q)$, obtained by the AXS method in Fig. 7.28 using the least-squares refining technique. The resultant structural parameters for Mo–Mo(Ni) pairs are listed in Table 7.8.

The structural unit in the 1.0 mol $l^{-1}$ $Na_2MoO_4$ solution is considered to be a distorted $MoO_6$ octahedron, similar to the one in the $MoO_3$ crystal. The distance of Mo–Mo pairs in the $Na_2MoO_4$ solution is 0.335 nm, as shown in Table 7.8. Then, the $MoO_6$ octahedra are likely to be connected by sharing their edges. Because of coincidence between the data of Mo–Mo pairs in the $Na_2MoO_4$ solution and the crystalline $Na_6Mo_7O_{24} \cdot 14H_2O$ case [97], it is quite likely that the local ordering structure consisting of the $MoO_6$ octahedra in solution is close to the cluster found in the crystal. Consequently, molybdenum ions in the 1.0 mol $l^{-1}$ $Na_2MoO_4$ solution form poly-molybdate ions as illustrated in Fig. 7.29a. Such poly-molybdate ions consist of seven $MoO_6$ octahedra, and they are connected by sharing their edges. This is consistent with the fact [98, 99] that molybdenum ions mainly form $Mo_7O_{24}^{6-}$ oxo-complexes consisting of seven $MoO_6$ octahedra in an acid molybdate aqueous solution.

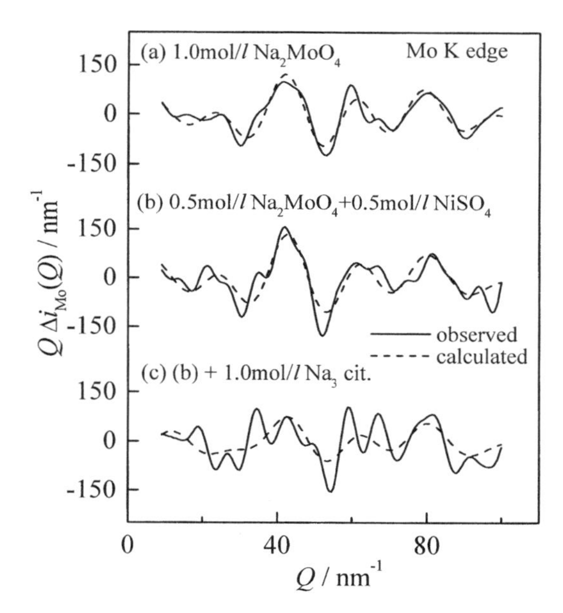


**Fig. 7.28. a–c** Environmental interference functions for Mo, $Q\Delta I_{\mathrm{Mo}}(Q)$, in aqueous solutions. **a** 1.0 mol $l^{-1}$ $Na_2MoO_4$, **b** 0.5 mol $l^{-1}$ $Na_2MoO_4$+0.5 mol $l^{-1}$ $NiSO_4$, **c** 0.5 mol $l^{-1}$ $Na_2MoO_4$+0.5 mol $l^{-1}$ $NiSO_4$+1.0 mol $l^{-1}$ sodium citrate dihydrate. *Dashed lines* denote the calculated values using the least-squares refining method

**Table 7.8.** Coordination numbers and distances of Mo–Mo pairs in 1.0 mol $l^{-1}$ $Na_2MoO_4$, 0.5 mol $l^{-1}$ $Na_2MoO_4$ + 0.5 mol $l^{-1}$ $NiSO_4$, and 0.5 mol $l^{-1}$ $Na_2MoO_4$ + 0.5 mol $l^{-1}$ $NiSO_4$ + 1.0 mol $l^{-1}$ $Na_3$ cit. aqueous solutions and $MoO_3$ crystal. Experimental errors: $N$: ±0.2; $r$: ±0.002 nm. Coordination numbers and distances of Mo–Mo and Mo–Ni pairs in crystalline $Na_6Mo_7O_{24} \cdot 14H_2O$ [97] and $Na_3(CrMo_6O_{24}H_6) \cdot 8H_2O$ [100] are also included. $Na_3$ cit.$Na_3C_6H_5O_7 \cdot 2H_2O$

| Sample | Mo–Mo | | Mo–Ni | |
|---|---|---|---|---|
| | $r$ (nm) | $N$ | $r$ (nm) | $N$ |
| 1.0 mol $l^{-1}$ $Na_2MoO_4$ sol. | 0.335 | 2.9 | – | – |
| 0.5 mol $l^{-1}$ $Na_2MoO_4$ + 0.5 mol $l^{-1}$ $NiSO_4$ sol. | 0.329 | 2.1 | 0.331 | 1.3 |
| $Na_6Mo_7O_{24} \cdot 14H_2O$ crystal | 0.333 | 3.1 | – | – |
| $Na_3(CrMo_6O_{24}H_6) \cdot 8H_2O$ crystal | 0.333 | 2.0 | 0.333 | 1.0 |

| | Mo–Mo(Ni) | |
|---|---|---|
| | $r$ / nm | $N$ |
| 0.5 mol $l^{-1}$ $Na_2MoO_4$ + 0.5 mol $l^{-1}$ $NiSO_4$ + 1.0 mol $l^{-1}$ $Na_3$ cit. sol. | 0.331 | 1.4 |

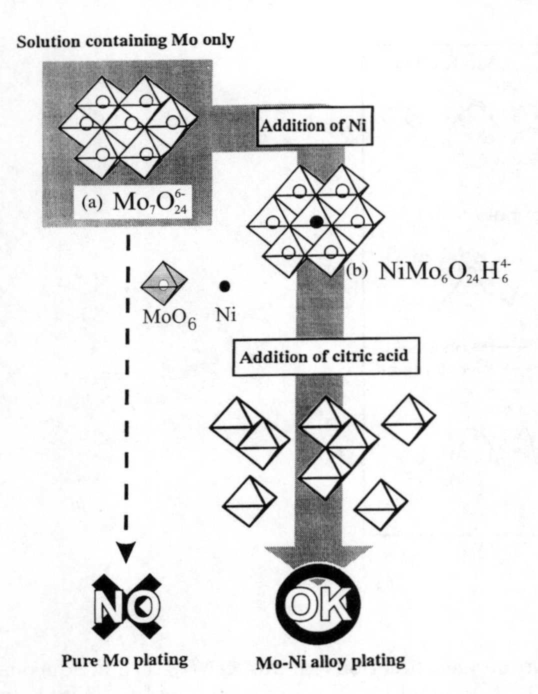


**Fig. 7.29.** Schematics of the structural models of poly-molybdate in aqueous solutions

The structural unit of the $MoO_6$ octahedron is unchanged in the presence of Ni ions, as is clearly seen in Table 7.7. In addition, we find in Table 7.8 that the coordination numbers and atomic distances of Mo–Mo and Mo–Ni pairs in the 0.5 mol $l^{-1}$ $Na_2MoO_4$ + 0.5 mol $l^{-1}$ $NiSO_4$ solution agree well with the values of the $Na_3(CrMo_6O_{24}H_6) \cdot 8H_2O$ crystal [100]. Thus, it is plausible that all the Mo ions and some of the Ni ions form a poly-molybdate ion in Fig. 7.29b which corresponds to the cluster found in the $Na_3(CrMo_6O_{24}H_6) \cdot 8H_2O$ crystal in spite of the difference between Ni and Cr. This result is also confirmed by the fact that the $Na_3(CrMo_6O_{24}H_6) \cdot 8H_2O$ single crystal was grown from an aqueous solution containing $Na_2MoO_4$ and $Cr(NO_3)_3$ at a 6:1 ratio with a pH of 4.5 [100].

As is easily seen in Fig. 7.27, the peak height due to Mo–Mo(Ni) pairs at about 0.33 nm is reduced in the solution containing citric ions. Although the total number of Mo and Ni around Mo at 0.33 nm is 3.4 in the 0.5 mol $l^{-1}$ $Na_2MoO_4$ + 0.5 mol $l^{-1}$ $NiSO_4$ solution, it becomes 1.4 by adding the 1.0 mol $l^{-1}$ sodium citrate dihydrate to the solution. This implies that clusters with a relatively large size are likely to present in the 0.5 mol $l^{-1}$ $Na_2MoO_4$ + 0.5 mol $l^{-1}$ $NiSO_4$ solution and only small ones are involved in the 0.5 mol $l^{-1}$ $Na_2MoO_4$ + 0.5 mol $l^{-1}$ $NiSO_4$ + 1.0 mol $l^{-1}$ sodium citrate dihydrate solution. Taking account of the decrease in the total coordination numbers around Mo in Table 7.8, it may safely be concluded that the large poly-molybdate ions in Fig. 7.29b are decomposed into small molybdate ions consisting of a few $MoO_6$ octahedra at most by adding citric ions to the 0.5 mol $l^{-1}$ $Na_2MoO_4$ + 0.5 mol $l^{-1}$ $NiSO_4$ solution. The mechanism of this decomposition process is probably explainable in the following way: Since a nickel ion is located in the center of the poly-molybdate ion and such nickel ions form complexes with citric ions, Mo and Ni ions cannot keep the large poly-molybdate ions any longer. Then, the large poly-molybdate ions containing the nickel ion are decomposed when citrates are added. It is worth mentioning that the small-angle X-ray scattering measurement from these solutions also supports the decomposition of the large poly-molybdate ions under the presence of citric ions [101, 102].

Molybdenum is electrodeposited only from solutions containing Mo, Ni and citric ions [103]. In such solutions, molybdenum ions exist as small molybdate ions which probably form citric complexes. Consequently, we consider that these small Mo–citrate complexes play a significant role in the mechanism of the induced co-deposition.

In conclusion, the potential capability of the AXS method to determine the structure around a metallic ion in aqueous solutions has clearly been demonstrated. The author maintains the view that the AXS method is very promising for structural characterization of various types of aqueous solutions, although its application to aqueous solutions is still limited.

## 7.4 Super-ionic Conducting Glasses

Super-ionic conductors (often called fast ion conductors) such as noble metal halides and noble metal-chalcogenides have received attention, because of their unusually high ionic conductivity (see, for example, [104]). For example, the tracer-diffusion coefficient of $Ag^+$ ions in noble metal chalcogenides is almost identical to the value of the self-diffusion coefficient in liquid silver. This particular property is of great importance when these electrolytes are considered for use as solid-state sensors. Super-ionic conducting glasses produced by rapid quenching from the melt have also been extensively studied by different techniques to delineate the relation between structure and high ionic conductivity [105, 106]. Then, some spectroscopic measurements using infrared, Raman and muclear magnetic resonance (NMR) and X-ray diffraction results are available. However, it is still difficult to provide conclusive evidence for structural origins in multi-component, non-crystalline materials from these data. Such question unsolved by conventional techniques can be answered by obtaining accurate information about the environmental structure around a specific element using the AXS method. For this reason, the identification of local atomic structure in multi-component, super-ionic conducting glass is described below, using the results of the $(CuI)_{0.3}(Cu_2O)_{0.35}(MoO_3)_{0.35}$ case [107] as an example. This is also based on the reason given below.

Recently much attention has been focused on the $Cu^+$ ion conducting glasses [108,109,110,111], because of their higher ionic conductivity in comparison with systems containing the $Ag^+$ ion. The CuI–$Cu_2O$–$MoO_3$ glass, first synthesized by Machida et al. [108,111], is one of the typical super-ionic conducting glasses showing high ionic conductivity of the order of 1 $\Omega^{-1}$ $m^{-1}$ at room temperature. Speculations on the structure of this glass system is based on data from infrared spectroscopy, invoking models consisting of $Cu^+$, $I^-$ and monomer $MoO_4$ ions. However, poly-molybdate complex structure formed by the $MoO_6$ unit has also been suggested [96,97], and the structural features of CuI–$Cu_2O$–$MoO_3$ glass have not been established yet.

Figure 7.30 shows the intensity profiles of the $(CuI)_{0.3}(Cu_2O)_{0.35}(MoO_3)_{0.35}$ glass measured at incident energies of 8.680 and 8.955 keV, below the Cu-K absorption edge (8.980 keV) [107]. The intensity difference, corresponding to the environmental structure around Cu in this glass, is illustrated at the bottom of this figure. Similar AXS measurements were carried out at incident energies of 19.702 and 19.977 below the Mo-K absorption edge (20.002 keV). The results are provided in Fig. 7.31. The intensity profiles essentially show typical features of glass structure having a broad peak at about 20 $nm^{-1}$. A pre-peak shoulder denoted by an arrow can also be appreciated in these profiles. It may be mentioning that such pre-peak shoulders are also observed in some super-ionic conducting glasses such as AgI–$Ag_2O$–$MoO_3$ [112] and AgI–$Ag_2O$–$P_2O_5$ [113,114] and they are attributed to density fluctuations of the $MoO_4$ chains or $PO_4$ chains [112,113,114,115]. Figure 7.32 gives the environmental RDFs obtained from the differential interference functions, $Q\Delta i_{\mathrm{Mo}}(Q)$ and $Q\Delta i_{\mathrm{Cu}}(Q)$, by Fourier transformation in $(CuI)_{0.3}(Cu_2O)_{0.35}(MoO_3)_{0.35}$ glass. The ordinary RDF estimated from the scattering profile at the single energy of 19.702 keV is also illustrated in this figure for comparison. It may be again stressed that a direct comparison among these three curves is only possible for the values on the abscissa, because different weighting factors are used on the ordinate.

In crystalline $Cu_2O$ and a number of oxo-cuprites [116,117], the Cu atom is typically found linearly coordinated by oxygen with a distance of 0.184 nm. On the other hand, the Mo atoms are known to exhibit octahedral coordination or tetrahedral coordination, as illustrated in Fig. 7.33 typical of $Mo^{6+}$ oxides [118], with the Mo–O distance ranging from 0.17 to 0.22 nm. For these reasons, it is difficult to identify the first peak in the ordinary RDF at around 0.19 nm with Cu–O and/or Mo–O pairs. In this subject, the AXS results can provide the answer by making available the scattering contrast of a desired element. For example, by comparing the ordinary RDF with two environmental RDFs, it is readily found that the correlation at a distance of 0.18 nm is completely lost in the environmental RDF for Cu. Thus, we can safely assign the first peak in the ordinary RDF to the Mo–O pair. The co-

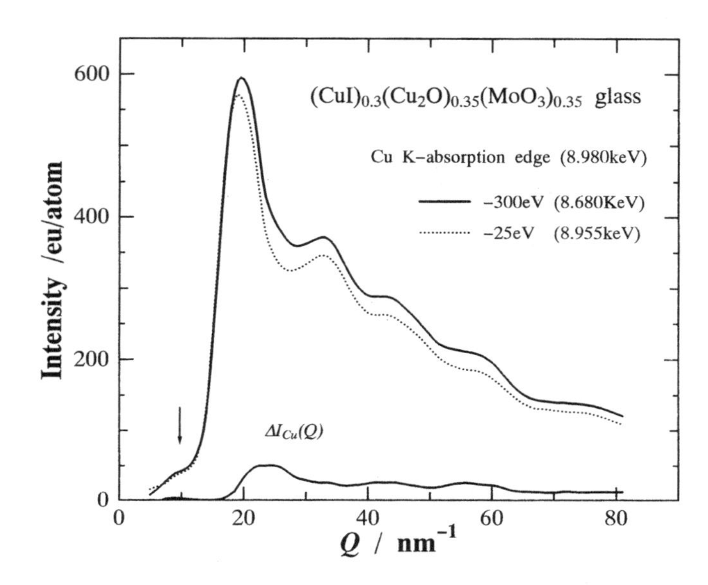


**Fig. 7.30.** Scattering intensity profiles of $(CuI)_{0.3}(Cu_2O)_{0.35}(MoO_3)_{0.35}$ glass measured at incident energies of 8.680 and 8.955 keV below the Cu-K absorption edge

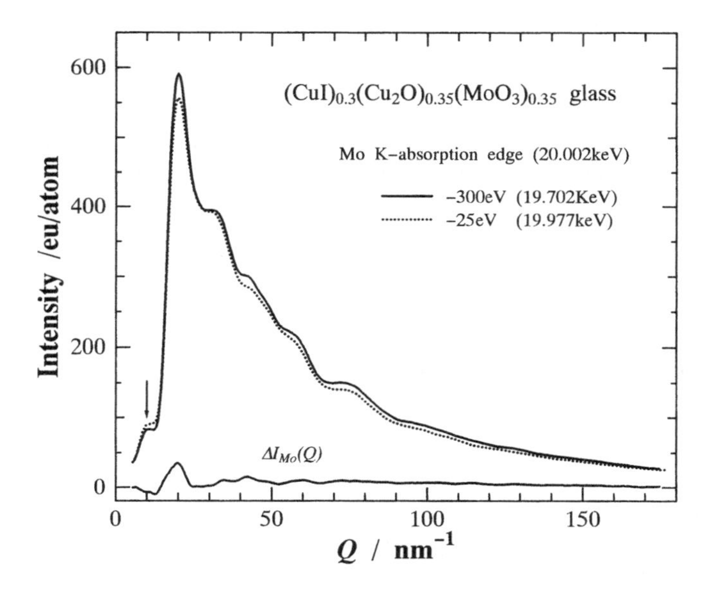


**Fig. 7.31.** Scattering intensity profiles of $(CuI)_{0.3}(Cu_2O)_{0.35}(MoO_3)_{0.35}$ glass measured at incident energies of 19.702 and 19.977 keV below the Mo-K absorption edge

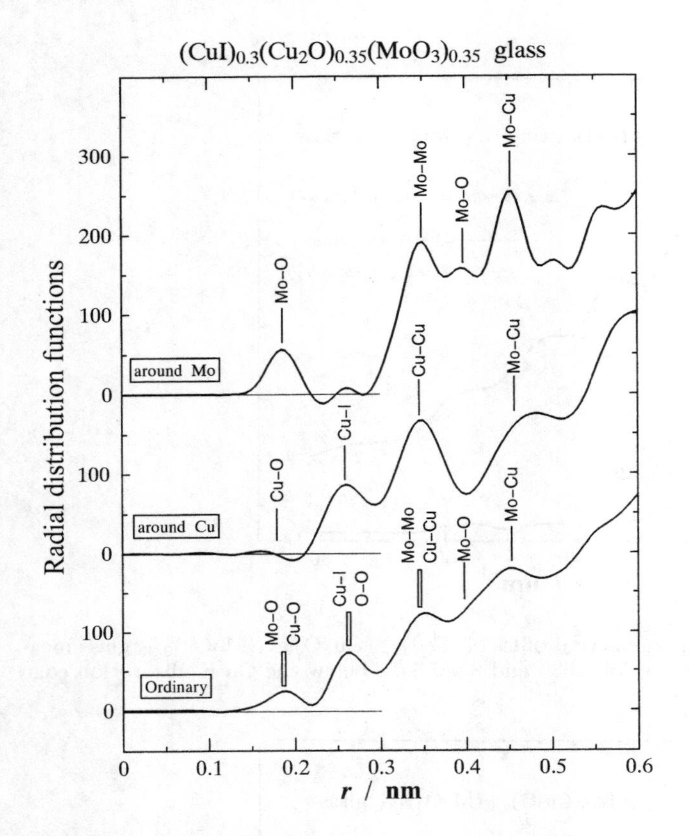


**Fig. 7.32.** Ordinary RDF and environmental RDFs around Mo and Cu. *Vertical lines* indicate the peak positions determined by the least-squares refining method

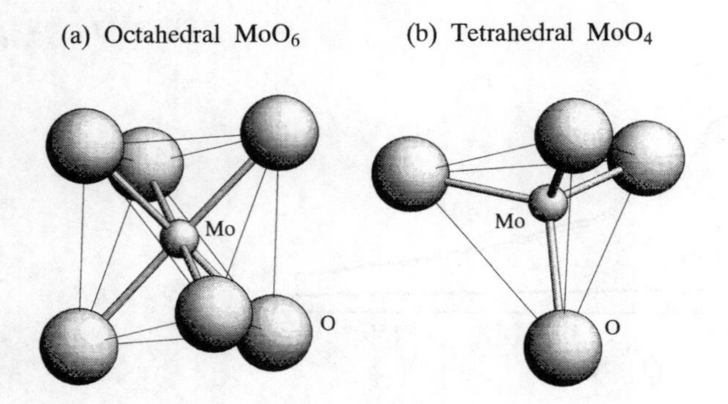


**Fig. 7.33.** Schematic of **a** an $MoO_6$ octahedral unit and **b** an $MoO_4$ tetrahedral unit

ordination number of the nearest-neighbor oxygens around Mo is estimated as 5.8 with conventional Gaussian fitting. The second peak observed in the ordinary RDF at around 0.26 nm almost completely disappears in the environmental RDF for Mo. This implies that the corresponding second peak is probably attributed to the Cu–I pair correlation, in view of the distance of the nearest-neighbor copper and iodine observed in both crystalline and molten CuI [31]. The coordination number of $I^-$ around $Cu^+$ estimated from this peak is estimated to be 1.2. However, the present AXS data for Cu is restricted to the wave vector, $Q$, range up to about 80 $nm^{-1}$, arising from the relatively low-energy absorption edge of Cu. Then, the reservation should be noted regarding the reliability of the absolute values of coordination number for Cu in this glass.

As already described in this chapter, the use of the least-squares refining method provides one way to minimize such a problem using the interference functions instead of RDFs. In this data processing, the distance and coordination number of interest in near-neighbor correlation are refined by the least-squares calculation [48, 119, 120] so as to reproduce three independent experimental data by the AXS method for Cu and Mo and ordinary diffraction. The resultant structural parameters are summarized in Table 7.9. The uncertainties, estimated from the variance–covariance matrix in the least-squares refining method [121], are also included in the table. The structural parameters listed reproduce three independent interference functions of $Q\Delta i_{\mathrm{Mo}}(Q)$, $Q\Delta i_{\mathrm{Cu}}(Q)$ and $Qi(Q)$, as shown in Fig. 7.34. This agreement is evidence that the present approach basically works well. Nevertheless, it should be kept in mind that this method is not a unique mathematical procedure, but the structural parameters in the near-neighbor region can be quantified with a much higher reliability, satisfying the necessary conditions best.

It is found in Table 7.9 that a Mo atom is surrounded with about six oxygens at 0.187 nm in the $(CuI)_{0.3}(Cu_2O)_{0.35}(MoO_3)_{0.35}$ glass. Namely, the AXS results clearly lead to the conclusion that the $MoO_6$ octahedral unit as shown in Fig. 7.33a is more likely to be present in the glass. This is no closer to the infrared spectroscopy data suggesting the presence of isolated $MoO_4$ tetrahedra in AgI–$Ag_2O$–$MoO_3$ glass [122, 123] than the results proposing the local structural units of $MoO_6$ octahedra [124]. In addition, distorted $MoO_6$ octahedral units are known to be predominant in most isopoly- and heteropoly-molybdates [125]. Nevertheless, it should be kept in mind that this concluding remark is not valid for all super-ionic conduction glasses containing the $MoO_3$ component. For example, the AXS results for the $(AgI)_{0.6}(Ag_2O \cdot MoO_3)_{0.4}$ glass [112] clearly identify that the most probable structural entity in this glass is the $MoO_4$ tetrahedral unit, and this is consistent with the conclusions derived earlier from infrared spectroscopy [122,123]. In other words, the controversy regarding the fundamental local unit structure in AgI–$Ag_2O$–$MoO_3$ glass drawn from two independent infrared studies [122, 124] can certainly be answered by the AXS method.

**Table 7.9.** Comparison of the structural parameters of the near-neighbor correlations for the $(CuI)_{0.3}(Cu_2O)_{0.35}(MoO_3)_{0.35}$ glass determined by the least-squares refining method

| Pairs | $r$ (nm) | $N$ |
|---|---|---|
| Mo–O | 0.187 ± 0.002 | 6.1 ± 0.4 |
| Mo–Mo | 0.350 ± 0.004 | 3.8 ± 0.5 |
| Mo–Mo | 0.394 ± 0.002 | 2.5 ± 0.2 |
| Cu–O | 0.186 ± 0.006 | 0.5 ± 0.4 |
| Cu–I | 0.261 ± 0.004 | 1.3 ± 0.4 |
| Cu–Cu | 0.341 ± 0.002 | 7.7 ± 0.3 |
| Mo–Cu | 0.449 ± 0.002 | 1.3 ± 0.6 |

The distance observed in the environmental RDF for Mo at 0.350 nm is allocated to the correlation of the Mo–Mo pair, enabling us to suggest a corner sharing linkage between the $MoO_6$ octahedra with a slight twist. The edge-shared or face-shared $MoO_6$ octahedra are ruled out, since these cases will contribute to a much shorter Mo–Mo distance of about 0.25 nm. The coordination number of the nearest Mo–Mo pairs is 3.9, suggesting that one $MoO_6$ octahedron is connected to four other octahedra on average with the corner-sharing linkage. The pre-peak shoulder observed at $Q = 10\ \text{nm}^{-1}$ in the intensity profiles would then originate from a density fluctuation due to such randomly connected $MoO_6$ units, where the correlation length is about 0.8 nm.

Table 7.9 also indicates that the average coordination number of nearest-neighboring Cu–O pairs is 0.5 with the distance of 0.186 nm. Then, some $Cu^+$ ions may be coordinated to the non-bridging oxygen. On the other hand, a $Cu^+$ ion appears to be surrounded with 1.3 iodine ions at 0.261 nm. This coordination number is distinctly smaller than the coordination number of the Cu–I pair in both crystalline [28, 126] and molten CuI [31], where four iodine ions are tetrahedrally coordinated around a $Cu^+$ ion. However, the value of 1.3 could be reduced to 4.3 $(= 1.3 \times 1.0(Cu_2O + CuI)/0.3(CuI))$ per $Cu^+$ ion generated from the CuI component. Any information on the origin of $Cu^+$ ions, whether from CuI or $Cu_2O$, cannot be identified from the present X-ray scattering data. Nevertheless, considering the experimental data on different mobilities for $Cu^+$ ions [108], the author is of the view that a fraction of the $Cu^+$ ions are surrounded by four iodine ions. The first nearest neighbor distance of Cu–Cu pairs is estimated to be 0.341 nm. This correlation can be found in molten CuI, whereas it is not recognized in crystalline CuI. Consequently, the local ordering structure of the added CuI transfers to the glassy oxide matrix, keeping the local environment similar to that of molten CuI.

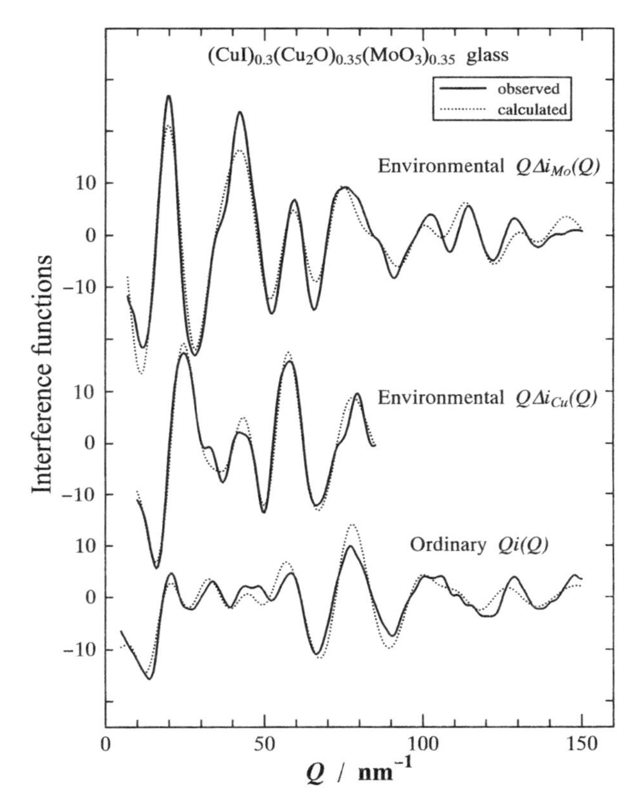


**Fig. 7.34.** The ordinary interference function, $Qi(Q)$, and the environmental interference functions, $Q\Delta i_{\mathrm{Mo}}(Q)$ and $Q\Delta i_{\mathrm{Cu}}(Q)$, of $(CuI)_{0.3}(Cu_2O)_{0.35}(MoO_3)_{0.35}$ glass. *Solid lines* correspond to the experimental data. *Dotted lines* denote values calculated by the least-squares refining method

On this particular glass of $(CuI)_{0.3}(Cu_2O)_{0.35}(MoO_3)_{0.35}$, XPS analysis has been used for characterizing the effects of argon-ion bombardment: and aging at room temperature on the surface properties. Copper and molybdenum ions in this glass are found to be significantly reduced by argon-ion bombardment: $Cu^{2+}$ to $Cu^0$ and $Mo^{6+}$ to $Mo^{4+}$. The chemical change is attributed to preferential sputtering of iodine and oxygen from the surface of the glass. The concentration of copper on the surface increases and that of molybdenum decreases with aging at room temperature. From aging curves, the diffusion coefficient of copper is estimated to be of the order of $10^{-23}\,\mathrm{m^2\,s^{-1}}$. It may also be noted that similar AXS analysis has been applied to the super-ionic conducting glass system of $GeSe_2$–$Ag_2Se$ [127] and $AgBr$–$Ag_2O$–$GeO_2$ [128]. Particularly, the environmental structure around all three glass components, for Ge, Se and Ag, are investigated in the $GeSe_2$–$Ag_2Se$ case.

## References

1. J.E. Enderby, D.M. North and P.A. Egelstaff: Philos. Mag., **14**, 961 (1966)
2. D.I. Page and I. Mika: J. Phys. C., **4**, 3034 (1971)
3. F.G. Edwards, J.E. Enderby, R.A. Howe and D.I. Page: J. Phys. C., **8**, 3483 (1975)
4. J.Y. Derrien and J. Dupuy: J. Physiq., **36**, 191 (1975)
5. J.Y. Derrien and J. Dupuy: Phys. Chem. Liquids, **5**, 71 (1976)
6. E.W.J. Mitchell, P.F.J. Poncet and R.J. Stewart: Philos. Mag., **34**, 721 (1976)
7. F.G. Edwards, R.A. Howe, J.E. Enderby and D.I. Page: J. Phys. C., **11**, 1053 (1978)
8. S. Biggin and J.E. Enderby: J. Phys. C., **14**, 3129 (1981)
9. S. Biggin and J.E. Enderby: J. Phys. C., **14**, 3577 (1981)
10. S. Biggin and J.E. Enderby: J. Phys. C., **15**, L305 (1982)
11. S. Eisenberg, J.F. Jal, J. Dupuy, P. Chieux and W. Knoll: Philos. Mag., A **46**, 195 (1982)
12. R.L. McGrevey and E.W.J. Mitchell: J. Phys. C., **15**, 5537 (1982)
13. J. Locke, R.L. McGreevy, S. Messoloras, E.W.J. Mitchell and R.J. Stewart: Philos. Mag., B **51**, 301 (1985)
14. K.F. Ludwig Jr., W.K. Warburton and L. Wilson: J. Chem. Phys., **87**, 604 (1987)
15. D.A. Allen and R.A. Howe: J. Phys. Condens. Matter, **4**, 6029 (1992)
16. P. Bodot: Acta Crystallogr., A **30**, 470 (1974)
17. M. Saito, C.Y. Park, K. Omote, K. Sugiyama and Y. Waseda: J. Phys. Soc. Jpn., **66**, 633 (1997)
18. M. Saito, S.C. Kang and Y. Waseda: Jpn. J.Appl. Phys. Suppl., **38**, 596 (1999)
19. R.G. Munro: Phys. Rev., B **25**, 5037 (1982)
20. R.L. McGreevy and L. Pusztai: Mol. Simul., **1**, 359 (1988)
21. R.L. McGreevy and L. Pusztai: Proc. Roy. Soc. Lond., A **430**, 241 (1990)
22. R.L. McGreevy: Nucl. Instrum. Meth. Phys. Res., A **354**, 1 (1995)
23. L. Pusztai and R.L. McGreevy:.J. Phys. Condens. Matter, **10**, 525 (1998)
24. A.K. Livesey and P.H. Gaskell: *Proc. of the Int. Conf. on Rapidly Quenched Metals, IVIV* (edited by T. Masumoto and K. Suzuki, Japan Institute of Metals, Sendai 1982) pp. 335
25. R.G. Stantson: *Numerical Methods for Science and Engineering* (Prentice-Hall, New York 1961)
26. C.L. Lawson and R.J. Hanson: *Solving Least-Square Problems* (Prentice-Hall, Englewood Cliffs, 1974)
27. G. Licheri, G. Navarra and S. Seatzu: J. Non-Cryst. Solids, **119**, 29 (1990)
28. W. Bührer and W. Hölg: Electrochim. Acta., **22**, 701 (1977)
29. M. Saito, S. Kang, K. Sugiyama and Y. Waseda: J.Phys. Soc. Jpn., **68**, 1932 (1999)
30. Y. Waseda, S. Kang, K. Sugiyama , M. Kimura and M. Saito: J. Phys. Condens. Matter, **12**, A195 (2000)
31. Y. Shirakawa, M. Saito, S. Tamaki, M. Inui and S. Takeda: J. Phys. Soc. Jpn., **60**, 2678 (1991)
32. M. Inui, S. Takeda, Y. Shirakawa, S. Tamaki, Y. Waseda and Y. Yamaguchi: J. Phys. Soc. Jpn., **60**, 3025 (1991)

33. M. Saito, C.Y. Park, K. Sugiyama and Y. Waseda: J.Phys. Soc. Jpn., **66**, 3120 (1997)
34. Y. Waseda, M. Saito, C. Park and K. Omote: J. Synchrotron Rd., **5**, 923 (1998)
35. P.H. Fuoss, P. Eisengerger, W.K. Warburton and A. Bienenstock: Phys. Rev. Lett., **46**, 1537 (1981)
36. S. Hosoya: Bull. Phys. Soc. Jpn., **25**, 110 and 228 (1970)
37. N.J. Shevchik: Philos. Mag., **35**, 805 and 1289 (1977)
38. A.K. Soper, G.W. Neilson, J.E. Enderby and R.A. Howe: J. Phys. C., **10**, 1793 (1977)
39. J.E. Enderby and G. Neilson: Rep. Prog. Phys., **44**, 593 (1981)
40. J.E. Enderby: J. Phys. C., **15**, 4609 (1982)
41. K. Takahashi, N. Mochida, H. Matsui, S. Takeuchi and Y. Gohshi: Yogyo-Kyokai-Shi, **84**, 482 (1976)
42. N. Mochida, T. Sekiya and A. Ohtsuka: Yogyo-Kyokai-Shi, **96**, 271 (1988)
43. K.Q. Lu, Y.Q. Zhao, L.C. Chang, Z.J. Shen, W.W. Huang and Y.F. Lin: Chin. Phys. Lett., **2**, 113 (1985)
44. Y. Shimizugawa, C.D. Yin, M. Okuno, H. Morikawa, F. Marumo,Y. Udagawa, N. Mochida and T. Sekiya: Yogyo-Kyokai-Shi, **95**, 418 (1987)
45. K. Sugiyama, Y. Waseda and M. Ashizuka: Mater. Trans. JIM, **32**, 1030 (1991)
46. E. Matsubara, K. Harada, Y. Waseda and M. Iwase: Z. Naturforsch., **43a**, 181 (1988)
47. Y. Waseda, E. Matsubara, K. Okuda, K. Omote, K. Tohji, S.N. Okuno and K. Inomata: J. Phys. Condens. Matter, **4**, 6355 (1992)
48. H.A.Levy, D.M. Danford and A.H.Narten: Oak Ridge National Laboratory Report, No. ORNL-3960 (ORNL, Oak Ridge 1966)
49. A.H. Narten, F. Vaslow and H.A. Levy: J. Chem. Phys., **58**, 5017 (1973)
50. A.H. Narten: J. Chem. Phys., **56**, 1905 (1972)
51. S.N. Okuno, S. Hashimoto, K. Inomata, S. Morimoto and A. Ito: J. Mag. Soc. Jpn., **14**, 213 (1990)
52. E. Matsubara, K. Okuda, Y. Waseda, S.N. Okuno and K. Inomata: Z. Naturforsch., **45a**, 1144 (1990)
53. P. Duwez, R.H. Williams and W. Klement: J. Appl. Phys., **31**, 36 (1960)
54. T. Masumoto and R. Maddin: Acta Meta., **19**, 191 (1971)
55. H.S. Chen, H.J. Leamy and M. Barmatz: J. Non-Cryst. Solids, **5**, 444 (1971)
56. A.Inoue, T. Zhang and T. Masumoto: Mater. Trans. JIM, **30**, 965 (1989)
57. A. Inoue: *Bulk Amorphous Alloys, Preparation and Fundamental Characteristics* (Trans. Tech., Uetkon-Zurich 1998)
58. N. Nishiyama and A. Inoue: Mater. Trans. JIM, **38**, 1531 (1996)
59. A. Inoue, N. Nishiyama and H. Kimura: Mater. Trans. JIM, **37**, 179 (1997)
60. A. Inoue: *Bulk Amorphous Alloys, Practical Characteristics and* (Applications, Trans. Tech. Uetkon-Zurich 1999)
61. T. Ikeda, E. Matsubara, Y. Waseda, A. Inoue, T. Chang and T. Masumoto: Mater. Trans. JIM, **36**, 1093 (1995)
62. P. Villars and L.D. Calvert (editors): *Peason's Handbook of Crystallographic Data for Intermetallic Phases, Vol.1-3* (American Society for Metals, Materials Park, Ohio 1985)
63. G. Petzow and G. Effenberg (editors): *Ternary Alloys, Vol.8* (VCH, New York, (1993)
64. E. Matsubara, T. Tamura, Y. Waseda, A. Inoue, T. Zhang and T. Masumoto: Mater. Trans. JIM, **33**, 873 (1992)
65. E. Matsubara, T. Tamura, Y. Waseda, T. Zhang, A. Inoue and T. Masumoto: J. Non-Cryst. Solids, **150**, 380 (1992)

66. E. Matsubara, K. Sugiyama, A.H. Shinohara, Y. Waseda, A. Inoue, T. Zhang and T. Masumoto: Mater. Sci. Eng., A **179/180**, 444 (1994)
67. A. Blesa and E. Matijevic: Adv. Colloid Interface Sci., **29**, 173 (1989)
68. Y. Waseda: *Novel Application of Anomalous X-ray Scattering for Structural Characterization of Disordered Materials* (Springer, Heidelberg, Berlin, New York 1984)
69. E. Matsubara and Y. Waseda: J. Phys. Condens. Matter, **1**, 8575 (1989)
70. K. Shinoda, E. Matsubara, M. Saito, Y. Waseda, T. Hirato and Y. Awakura: Z. Naturforsch., **52a**, 855 (1997)
71. K.F. Ludwig Jr., W.K. Warburton and A. Fontaine: J. Chem. Phys., **87**, 620 (1987)
72. A.H. Narten and H.A. Levy: Science **160** 447 (1969)
73. J.E. Enderby and G.W. Neilson: Adv. Phys., **29**, 323 (1980)
74. P. Lagrade, A. Fontaine, D. Raoux, A. Sadoc and P. Migliardo: J. Chem. Phys., **72**, 3061 (1980)
75. D.L. Wertz and R.F. Kruh: J. Chem. Phys., **43**, 2163 (1965)
76. D.L. Wertz and R.F. Kruh: J. Chem. Phys., **50**, 4313 (1969)
77. H. Ohtaki, T. Yamaguchi and M. Maeda: Bull. Chem. Soc. Jpn., **49**, 701 (1976)
78. R. Caminiti and P. Cucca: Chem. Phys. Lett., **89**, 110 (1982)
79. P. Dreier and P. Rabe: J. Phys., **47**, C8-809 (1986)
80. E. Matsubara, K. Okuda and Y. Waseda, J. Phys.: Condens. Matter, **2**, 9133 (1990)
81. S. Biggin, J.E. Enderby, R.L. Hahn and A.H. Narten: J. Phys. Chem., **88**, 3634 (1984)
82. V.Q. Kinh, E. Chassaing and M. Saurat: Elecrodep. Sur. Treat., **3**, 205 (1975)
83. K. WiKeil and J. Osteryoung: J. Appl.Electrochem., **22**, 506 (1992)
84. I.A. Raj: J. Mater. Sci., **28**, 4375 (1993)
85. A. Brenner: *Electrodeposition of Alloys 2* (Academic Press, New York 1963) Chap. 34
86. D.W. Ernst and M.L. Holt: J. Electrochem. Soc., **105**, 686 (1958)
87. S. Rengakuji, Y. Nakamura, M. Inoue, K. Nishibe and H. Imanaga: Denki Kagaku, **59**, 885 (1991)
88. H. Fukushima, T. Akiyama, S. Akagi and K. Higashi: Mater. Trans. JIM, **20**, 358 (1979)
89. J. Podlaha, M. Matlosz and D. Landolt: J. Electrochem. Soc., **140**, L149 (1993)
90. E.J. Podlaha and D. Landolt: J. Electrochem. Soc., **143**, 885 (1993)
91. K. Higashi and H. Fukushima: J. Metal. Finish. Soc. Jpn., **24**, 486 (1973)
92. C.C. Nee, W. Kim and R. Weil:J. Electrochem. Soc., **135**, 1100 (1988)
93. B. Yuan, K. Liu and M. Kowaka: J. Metal. Finish. Soc. Jpn., **42**, 639 (1991)
94. S. Yao and M. Kowaka: Metal. Finish. Soc. Jpn., **39**, 736 (1988)
95. T. Aliyama, H. Fukushima, F. Yuse, T. Tsuru and Y. Tomokiyo: Metal. Finish. Soc. Jpn., **46**, 1133 (1995)
96. L. Kihlborg: Arkiv Kemi., **21**, 357 (1963)
97. K. Sjobom and B. Hedman: Acta Chem. Scand., **27**, 3673 (1973)
98. G. Johansson, L. Pettersson and N. Ingri: Acta Chem. Scand., A **28**, 1119 (1974)
99. F.A. Cotton and G. Wilkinson: *Advanced Inorganic Chemistry* (5th ed., Wiley-Interscience, New York 1988) pp. 817
100. A. Perloff: Inorg. Chem., **9**, 2228 (1970)
101. E. Uekawa, K. Murase, E. Matsubara, T. Hirato and Y. Awakura: J. Electrochem, Soc., **145**, 523 (1998)

102. K. Shinoda, E. Matsubara, Y. Waseda, E. Uekawa, K. Murase, T. Hirato and Y. Awakura: Hyomen-Gijutsu, **49**, 1115 (1998)
103. T. Watanabe, T. Naoe, A. Mitsuo and S. Katsumata: J. Finish. Soc. Jpn., **40**, 458 (1989)
104. P. Hagenmüller and W. van Goal (editors): *Solid Electrolytes* (Academic Press, New York 1978)
105. J.L. Souquet: Solid State Ionics, **5**, 77 (1981)
106. T. Minami: J. Non-Cryst. Solids, **56**, 15 (1983)
107. M. Saito, K. Sugiyama, E. Matsubara, K.T. Jacob and Y. Waseda: Mater. Trans. JIM, **36**, 1434 (1995)
108. N. Machida, M. Chusho and T. Minami: J. Non-Cryst. Solids, **101**, 101 (1988)
109. C. Liu and C.A. Angell: Solid State Ionics, **13**, 105 (1984)
110. T. Minami and N. Machida: Mater. Chem. Phys., **23**, 51 (1989)
111. N. Machida and T. Minami: J. Am. Ceram. Soc., **71**, 784 (1988)
112. M. Saito, K.T. Jacob and Y. Waseda: High Temp. Mater. Process, **18**, 21 (1999)
113. M. Tachez, R. Mercier, J.P. Malugani and P. Chieux: Solid State Ionics, **25**, 2772 (1984)
114. H. Takahashi, E. Matsubara and Y. Waseda: J. Mater. Sci., **29**, 2536 (1994)
115. L. Börjesson, R.L. McGreevy and W.S. Howells: Philos. Mag., B **65**, 261 (1992)
116. K. Hestermann and R. Hoppe: Z. Anorg. Allg. Chem., **360**, 113 (1968)
117. H-M. Higeon, M. Zanne, C. Gleitzer and A. Courtois: J. Solid State Chem., **16**, 325 (1976)
118. J.B. Goodenough: *Proc. of the Climax 4th Int. Conf. on the Chem. and Uses of Molybdenum*, ed. by H.F. Barry and P.C.H. Mitchel (Climax Molybdenum, Ann Arbor 1982)
119. P.A. Johnson, A.C. Wright and R.N. Sinclair: J. Non-Cryst. Solids, **50**, 281 (1982)
120. E. Matsubara, K. Sugiyama, Y. Waseda, M. Ashizuka and E. Ishida: J. Mater. Sci. Lett., **9**, 14 (1990)
121. Y. Waseda, E. Matsubara, K. Sugiyama, I.K. Suh, T. Kawazoe, O. Kasu, M. Ashizuka and E. Ishida: Sci. Rep. Res. Inst. Tohoku Univ., A **35**, 19 (1990)
122. T. Minami and M. Tanaka: J. Non-Cryst. Solids, **38/39**, 289 (1980)
123. S. Hemalatha, P.R. Sarode and K.J. Rao: J. Non-Cryst. Solids, **54**, 313 (1983)
124. A. Rajalakshmi, M. Seshasayee, G. Aravamudan, T. Yamaguchi, M. Nomura and H. Ohtaki: J. Phys. Soc. Jpn., **59**, 1252 (1990)
125. A.G. Wells: *Structural Inorganic Chemistry* (Clarendon Oxford 1975)
126. K. Funke: Progr. Solid State Chem., **11**, 345 (1976)
127. J.D. Westwood, P. Georgopoulos and D.M. Whitmore: J. Non-Cryst. Solids, **107**, 88 (1988)
128. S. Kang, M. Saito, K.T. Jacob and Y. Waseda: Sci. Tech. Adv. Mater., **1**, 37 (2000)

# 8. Anomalous Small-Angle X-ray Scattering

Small-angle X-ray scattering (hereafter referred to as SAXS) [1] has been widely used for characterizing the atomic-scale structure of materials, because the SAXS data enable us to offer many important microstructure parameters, such as the particle volume and particle shape producing the structural inhomogeneity in a sample. Since the interpretation of the SAXS data, in principle, depends on the models used for theoretical calculation of the intensity, it is frequently found that more than two kinds of models can fit equally well the experimental data. The use of the AXS method is one way to overcome such experimental difficulties, by making available sufficient atomic sensitivity arising from the anomalous dispersion effect of a specific element. For this reason, the harmonic combination of SAXS and the anomalous dispersion effect has emerged as a new and powerful method for characterizing the inhomogeneities in materials at a microscopic level. The usefulness of this novel application will be described in this chapter with some selected examples.

## 8.1 Nature of Guinier–Preston Zones

The feasibility of the anomalous SAXS method has been tested for determining the nature of Guinier–Preston(GP) zones [1] of phase-separated Al–Zn alloys by comparing scattering intensities at different energies of the incident X-rays close to the absorption edge of Zn [2]. The Al–Zn alloy is a good example, because metallurgists have long accepted GP zones as one of the essential subjects for aluminum-base alloys, the phase relation of this binary alloy is well known and a large amount of the conventional SAXS data are available.

Figure 8.1 shows the absolute intensity data scattered by GP zones in Al–Zn alloys as a function of the wave vector measured at energies below the Zn absorption edge. The inset provides the SAXS data of the Al–Ag alloy as a reference sample where no anomalous dispersion effect is detected. These measured intensity data were corrected for beam decay, detector response, absorption, parasitic scattering and Laue background. As easily seen in the results of Fig. 8.1, strong energy dependence, in other words, the anomalous

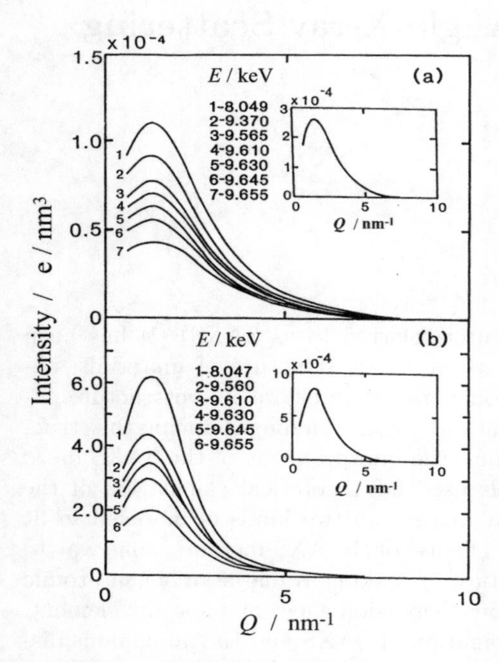


**Fig. 8.1. a,b** Absolute intensities scattered by the GP zones in Al–Zn alloys at various energies below the Zn absorption edge at 9.660 keV. **a** Al–4.4at.-%Zn at several energies. *Inset*: scattering from the Al–1.1at.-%Ag control sample at the same energies. **b** Al–6.7at.-%Zn at several energies. *Inset*: scattering from the Al–2.4at.-%Ag at the same energies [2]

scattering effect of Zn, explicitly appears in the SAXS intensities for two Al–Zn alloys. This contrasts with the Al–Ag alloy case. The anomalous SAXS method has been extended to Al–Zn–Ag ternary alloys by Lyon and his colleagues [3, 4], and variation in SAXS intensity is provided in Fig. 8.2 using the results for the $Al_{82}Zn_9Ag_9$ alloy aged at 398 K for 600 s. When the incident X-ray energy increases from 8.50 to 9.658 keV, the SAXS intensity also appears to monotonically decrease due to the anomalous dispersion effect of Zn.

The particle size of precipitates and their size dynamics can be estimated from conventional SAXS analysis such as Guinier plots, and the present Al–Zn–Ag alloy case is found to be consistent with the so-called "two-phases unmixing model". This model is expressed by precipitates with a sharp interface embedded in a matrix, and both phases have uniform concentrations, $c_i^{\mathrm{P}}$ and $c_i^{\mathrm{M}}$, respectively. In this case, all partial structure factors are likely to have the same shape factor, $S_{\mathrm{PM}}(Q)$, and the X-ray scattering intensity may be approximated as follows:

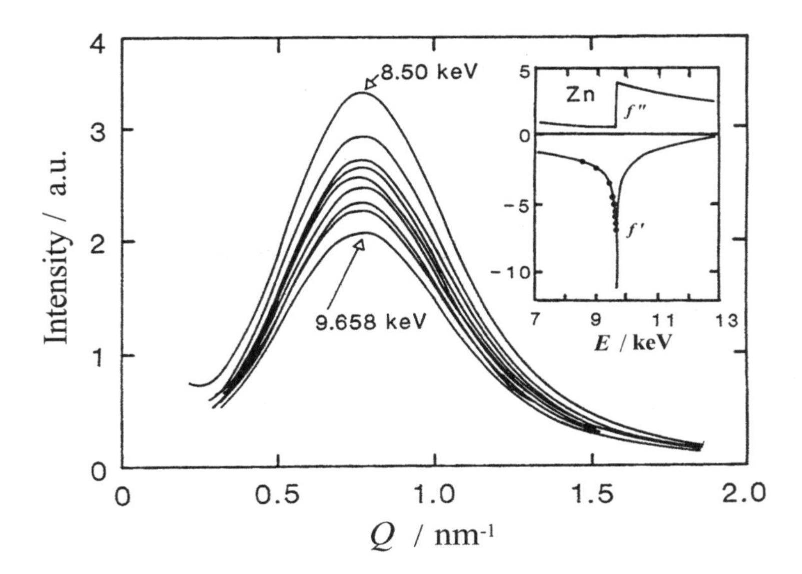


**Fig. 8.2.** Variation of SAXS intensity for the Al–9at.-%–Zn–9at.-%–Ag alloy aged at 398 K for 600 s. Upper to lower curves were recorded at X-ray energies of 8.50, 9.00, 9.40, 9.56, 9.625, 9.64, 9.651 and 9.65 keV, respectively [4]. All energies are located below the absorption edge

$$I(Q) = \left( \sum_{i,j}^{2} \Delta c_i \Delta c_j F_i F_j \right)^2 S_{\mathrm{PM}}(Q), \tag{8.1}$$

where $\Delta c_i = c_i^{\mathrm{P}} - c_i^{\mathrm{M}}$ and $F_i = f_i - f_\circ$. A simple way to analyze the anomalous scattering data of the Al–Zn–Ag alloy is to rewrite (8.1) as $(\Delta c_{\mathrm{Zn}} F_{\mathrm{Zn}} + \Delta c_{\mathrm{Ag}} F_{\mathrm{Ag}})^2 S_{\mathrm{PM}}(Q)$ and then, when plotting the relation between $\sqrt{I(Q)}$ and $F_{Zn}$ as a function of energy, a straight line could be predicted; its slope would provide the ratio of $\Delta c_{\mathrm{Ag}} / \Delta c_{\mathrm{Zn}}$.

For example, during growth and coarsening of GP zones in the $Al_{82}Zn_9Ag_9$ alloy, the precipitate particle composition denoted by the ratio of Ag and Zn remains fairly constant, of the order of 1.10. This implies a rather balanced diffusion of Zn and Ag towards the GP zones. On the other hand, the ratio of Ag and Zn drastically changed from 0.475 to 0.275 in the $Al_{82}Zn_9Ag_9$ alloy, as the precipitate particles transform from GP zones to the stable $\varepsilon'$ precipitate particles [3, 4]. It may be worth mentioning that a rotation of the corresponding tie line is also suggested in this un-mixing process. These interesting concluding remarks cannot be extracted from the conventional SAXS data alone.

The anomalous SAXS method at the single edge was applied to other metallic alloys such as Fe–Ni–X–Mo (X = Mn or Co) alloys [5, 6] and Ti–X–

Mo (X = Al or Nb) alloys [7,8]. However, it should rather be stressed that the AXS measurements using energies in the close vicinity of more than two edges are strongly requested for obtaining further information which makes our understanding of structural inhomogeneity better.

## 8.2 Composition Modulation in Amorphous Alloys

A large number of amorphous alloys have been produced by several techniques such as rapid quenching from the melt, vapor deposition and mechanical grinding. Since some of these new alloys have technological potential in applications such as soft magnetic devices, amorphous alloys can no longer be considered as a laboratory curiosities. The distinguished feature of amorphous alloys is rather characterized by the lack of crystal-like atomic periodicity. Since such particular atomic non-periodicity is not completely stable, many physical properties of amorphous alloys often significantly depend on the thermal history. It is suggested that the conventional SAXS, as well as wide-angle X-ray scattering, cannot give a well-defined conclusion with respect to the origin of fine-scale composition modulation induced from density fluctuation on a scale of tens to hundreds of nanometers, both in the crystalline and non-crystalline states. However, the anomalous SAXS method again holds promise in reducing this difficulty by obtaining the sufficiently significant scattering contrast of a desired element due to the anomalous dispersion effect from a specific element.

Metal-germanium amorphous films can be prepared in a wide concentration range, where the resultant amorphous films are semiconductive at low metal concentration and become metallic when the metal component reaches 10–25 at.-%. In a higher metal concentration range, for example, superconductivity appears at 13 at.-% Mo in the Mo–Ge system, and ferromagnetism is detected at 40 at.-% Fe in the Fe–Ge system. The anomalous SAXS measurements for amorphous metal-germanium alloys have been made in order to determine "whether these interesting properties occur via phase separation or through homogeneous alloy formation" [9,10].

Figure 8.3 shows the variation of the anomalous SAXS intensity of the amorphous $Fe_{12}Ge_{88}$ alloy measured at energies below the absorption edges of Fe and Ge. The SAXS intensity apparently decreases as the incident X-ray energy approaches the absorption edge of Fe from the lower-energy side. In contrast, the SAXS intensity remains almost constant when changing in the close vicinity of the absorption edge of Ge. Rice et al. [9] confirmed that similar variation in the anomalous SAXS intensity is also obtained in another two cases: amorphous $W_8Ge_{92}$ and $Mo_8Ge_{92}$ alloys. These results clearly suggest that germanium does not play a significant role in the intensity of the small-angle region by its uniform distribution, whereas the measured SAXS intensity should be attributed mainly to fluctuations in the electron density of metal atoms (Fe, W and Mo) in these amorphous alloy films. It is

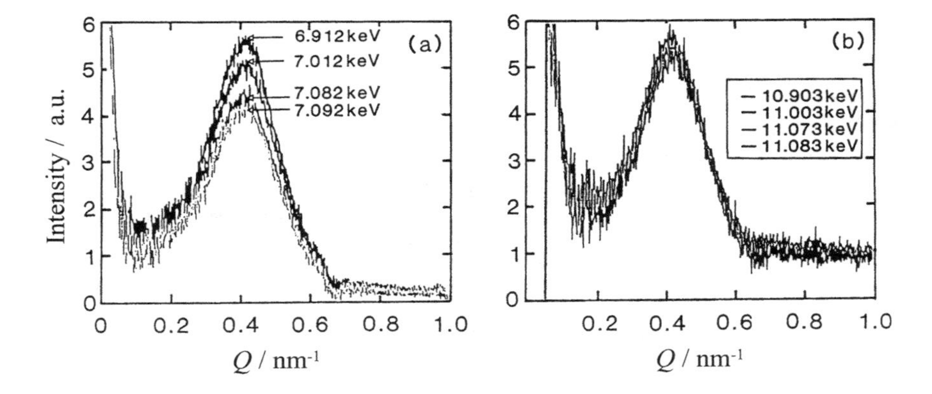


**Fig. 8.3. a,b** The intensity of scattered radiation at incident energies of 200, 100, 30 and 20 eV below the **a** Fe- and **b** Ge-K absorption edges in amorphous $Fe_{12}Ge_{88}$ [9]

interesting to note that all three amorphous metal-germanium film samples show a peak in the SAXS intensity near 0.4 $nm^{-1}$, which corresponds to a spatial correlation length of the order of 2 nm.

As shown in Fig. 8.4 using the results of an amorphous $Fe_x$–$Ge_{100-x}$ alloy as an example, the peak position in the SAXS intensity shifts towards smaller values of the wave vector with an increasing metallic component. This corresponds to the variation in correlation length of the compositional fluctuation from 2 to 14 nm. It may also be cited that the peak intensity increases with increasing metallic component. Thus, it can be concluded from these systematic results that very-fine-scale composition modulation is quite likely to occur in the low metal concentration range of amorphous metal-germanium alloys, and its size scale is increased as the metallic component increases. Coupled with the data for number densities of crystalline Ge and $MGe_2$ [9, 10], phase separation of amorphous metal-germanium alloys into an amorphous Ge matrix and an amorphous $MGe_2$-like substance is also proposed in the composition range of $0 < x < 33$.

In 1993, two stages of glass transition temperatures, as well as a wide super-cooled liquid region, were found in the amorphous $Zr_{33}Y_{27}Al_{15}Ni_{25}$ alloy [11]. The anomalous AXS measurement was carried out in order to identify the origin of this particular behavior, although the structural inhomogeneity attributed to the insoluble character of Y and Zr was qualitatively proposed. As shown in Fig. 8.5, a maximum of the SAXS intensity of the amorphous $Zr_{33}Y_{27}Al_{15}Ni_{25}$ alloy is observed at a wave vector of 0.65 $nm^{-1}$, indicating a spatial correlation length of the order of 10 nm [12]. In the case of a sample annealed at 773 K for more than 600 s, a drastic increase in intensity is detected near the angular origin. This is consistent with the results related to the crystallization behavior of the $Al_3Zr_5$ phase [11]. For this reason, the sample annealed at 773 K for 300 s is chosen as the best condition for studying the structural inhomogeneity in this particular amorphous alloy.

Figure 8.6 provides the energy dependence of the SAXS intensity for the amorphous $Zr_{33}Y_{27}Al_{15}Ni_{25}$ alloy measured at energies close to the K edge of Ni, Y and Zr [12]. The inset corresponds to the SAXS data of Al–6.7at.-% Zn alloy as a reference sample in which no anomalous dispersion effect is detected when measured at the same energy region. This supports the proposition that the present AXS measurements work well. Although the positional resolution of anomalous SAXS signals might be insufficient, the anomalous dispersion effect is clearly confirmed in these three results by detecting the remarkable decrease in SAXS intensity for Ni and Zr when the incident X-ray energy approaches the respective K absorption edge. On the contrary, the SAXS intensity for Y increases as the absorption edge of Y is approached from the lower-energy side. The variation in intensity of the anomalous SAXS peak for the amorphous $Zr_{33}Y_{27}Al_{15}Ni_{25}$ alloy between the two results obtained at the two energies of –30 and –300 eV (– 200 eV for the Zr case) was estimated as +18, –21 and +50% for Ni, Y and Zr, respectively.

It is quite interesting to reveal a possible origin for the structural inhomogeneity of this particular amorphous alloy related to the SAXS peak at $Q = 0.65\ \mathrm{nm}^{-1}$, by considering the interference effect due to fluctuation in composition and density attributed to a certain precipitate such as $Al_3Zr_5$. The essential points for this data processing are described below.

Assuming that the present amorphous alloy annealed at 773 K consists of two phases, for example, a crystalline $Al_3Zr_5$ precipitate and the amorphous matrix, the total integrated intensity, $Q_\circ$, is proportional to the square of the electron density difference between two phases of $\Delta\rho$, as follows [13]:

$$Q_\circ = \frac{1}{2\pi^2}\int_{Q_{\min}}^{Q_c} Q^2 I(Q)\mathrm{d}Q = (\Delta\rho)^2, \tag{8.2}$$

where $I(Q)$ is the coherent scattering intensity and $Q_c$ and $Q_{\min}$ are the experimental cut off value and the minimum value for the wave vector, respectively. The volume of fraction of the inhomogenious region is considered to be constant for a given sample. Thus, the integrated scattering intensity could be expressed by the square of the average electron density difference. Furthermore, the integrated intensity may also be proportional to the quantity of $[(\Delta<f>)^2]$, corresponding to the difference in the average atomic scattering factors between two phases.

The intensity variations at the three absorption edges were calculated using the values of the anomalous dispersion factors with the measured density of 5.62 $\mathrm{Mg\,m^{-3}}$ for the amorphous $Zr_{33}Y_{27}Al_{15}Ni_{25}$ alloy. The calculated variation in intensity at the Y edge decreases when the incident X-ray energy approaches the absorption edge from the lower-energy side, and this is inconsistent with the experimental result. Similarly, the calculated variation at the Zr edge also contradicts the experimental result. For this reason, the

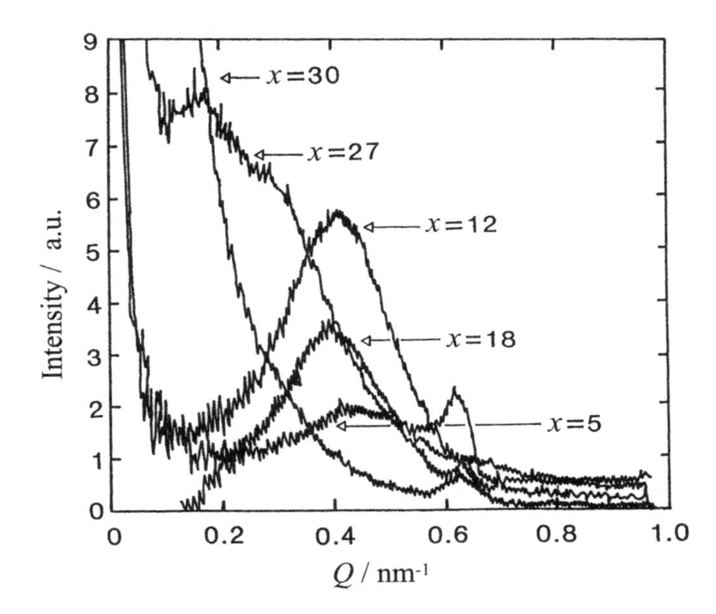


**Fig. 8.4.** The variation of peak intensity and position in amorphous $Fe_xGe_{100-x}$ for $x$ values of 5, 12, 18, 27 and 30. The small narrow peak near $Q = 0.65\ nm^{-1}$, which appears in several samples, is due to imperfect substraction of the kapton substrate [9]

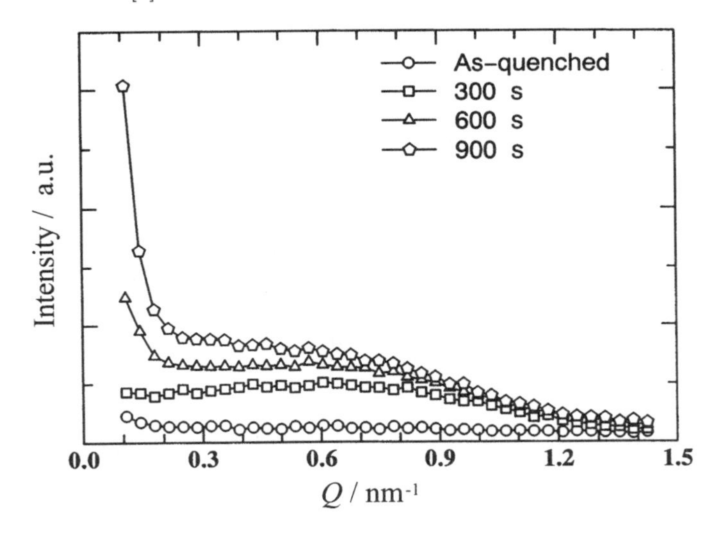


**Fig. 8.5.** SAXS intensity patterns of the amorphous $Zr_{33}Y_{27}Al_{15}Ni_{25}$ alloy annealed at 773 K for various lengths of time [12]

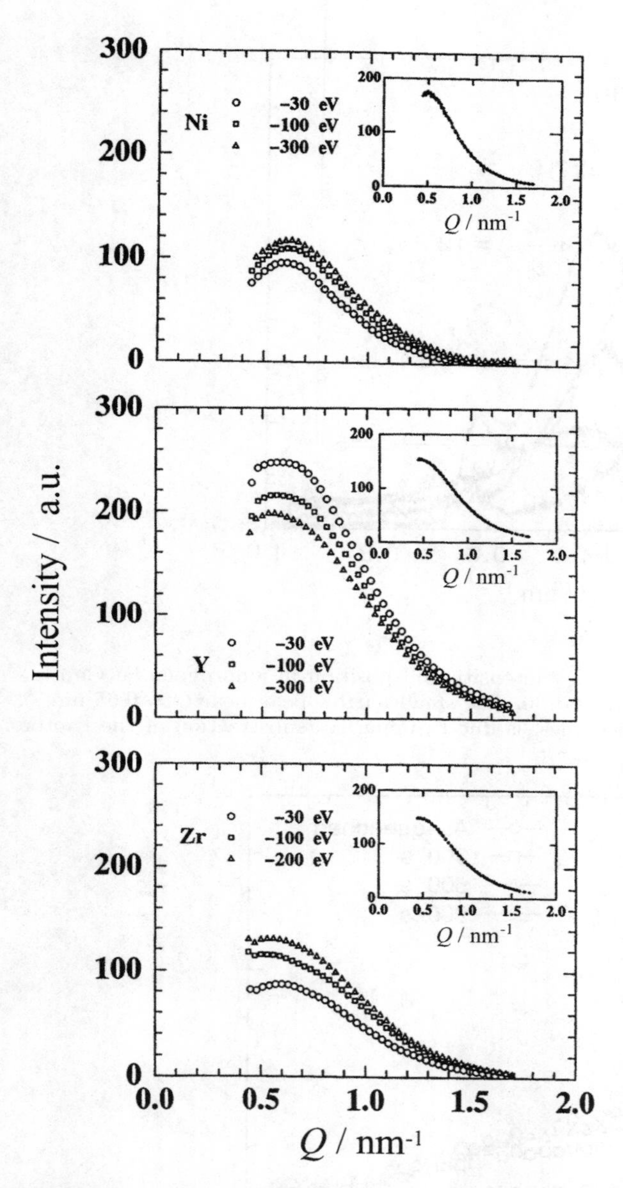


**Fig. 8.6.** Energy dependence of the SAXS intensities measured at energies near the K absorption edge of Ni, Y and Zr for the amorphous $Zr_{33}Y_{27}Al_{15}Ni_{25}$ alloy annealed at 773 K for 300 s [12]

crystalline $Al_3Zr_5$ precipitate is not accepted as a possible origin for the structural inhomogeneity in this amorphous alloy. In other words, the structural inhomogeneity, giving rise to the SAXS signal, should be linked to another combination.

Let us recall the local structural change of this particular amorphous alloy induced by annealing. The wide-angle AXS results for the amorphous $Zr_{33}Y_{27}Al_{15}Ni_{25}$ alloy [14,15] propose that the chemical fluctuation of Y-rich precipitates is likely to exist, and this leads us to imagine that a crystalline phase such as $Y_{60}Al_{40}$ or an amorphous phase such as $Y_{60}Al_{15}Ni_{25}$ is segregated from the amorphous $Zr_{33}Y_{27}Al_{15}Ni_{25}$ matrix.

The intensity variation at the three absorption edges were again calculated with the density values of 4.36 $Mg\,m^{-3}$ for crystalline $Y_{60}Al_{40}$ and 4.95 $Mg\,m^{-3}$ for amorphous $Y_{60}Al_{15}Ni_{25}$, respectively. The results are summarized in Table 8.1 together with the experimental data [12]. Present combination provides the correct sign for the intensity variation at three absorption edges, and it should be stressed that the $Y_{60}Al_{15}Ni_{25}$ case agrees well with the experimental results, although the quantitative agreement between calculation and experiment at the Ni edge is poor. Further consideration should be made using the precise value of the atomic volume of the precipitate, as well as the accurate integrated intensity. However, such further work is beyond the limit of the present experimental uncertainty. Nevertheless, the usefulness of the anomalous SAXS method is clearly demonstrated by obtaining the possible origin for the structural inhomogeneity in the amorphous $Zr_{33}Y_{27}Al_{15}Ni_{25}$ alloy. It is undoubtedly difficult to solve using the conventional SAXS data alone.

The anomalous SAXS method has also been successfully applied to amorphous CuTi [13] and (Tb,Gd)Cu alloys [16]. It is also worth noting from the anomalous SAXS data that the amorphous $(Fe,Mn)_{35}Y_{65}$ alloy was found to be represented by segregation attributed to the concentration fluctuations and the partial atomic volume ratio of (Fe,Mn) and Y [17].

**Table 8.1.** The variation of the scattered intensity when the incident energy leaves the absorption edge. $Y_{60}Y_{40}/Zr_{33}$: segregation of $Y_{60}Al_{40}$ with the matrix of amorphous $Zr_{33}Y_{27}Al_{15}Ni_{25}$

| | Ni | Y | Zr |
|---|---|---|---|
| Experimental | +18 | −21 | +50 |
| $Y_{60}Al_{40}/Ze_{33}$ | +16 | −9 | +18 |
| $Y_{60}Al_{15}Ni_{25}$ / $Zr_{33}$ | +4 | −21 | +59 |

## 8.3 Concentration Profile and Specific Volume in Multi-layered Thin Films

Recently, new synthetic multi-layered films have received much attention, because of their scientific and technological potential in areas such as the soft X-ray monochromator, where elements with low and high atomic numbers are alternately stacked with a certain periodic thickness in order to obtain an effective scattering power. A magnetic multilayer is also one of the new applications. Particularly, the gigantic magnetoresistance of a Co/Cu multilayer has intensively been studied. These metallic superlattice films also show the antiferro- and ferro-magnetic oscillation behavior with an increase in the thickness of the layers of the non-magnetic component [18, 19]. Although the origin of these peculiar features of multi-layered films is not well identified yet, their periodic and interfacial structures play an important role in achieving optimum performance with these new advanced materials [20, 21].

Diffraction peaks observed in the small-angle region provide us information about the periodicity, the thickness of individual layers and the concentration profile across the interface. Their peak intensities are known to be proportional to the square of the difference between the average X-ray atomic scattering factors of the corresponding layers. On the other hand, some interesting multi-layered films consist of elements with close atomic numbers, such as Mn, Fe, Co, Ni and Cu; in such a case, weak diffraction peaks are expected only. With regard to this subject, the AXS method is quite effective by making available sufficient atomic sensitivity, even for the case containing next-neighbor elements in the periodic table [22, 23]. When coupled with the differential anomalous SAXS, a new method for analyzing the periodic structure of a multilayer has also been proposed in order to evaluate an accurate peak profile even for the higher-order peaks [24]. The essential points of this new method are described below, using the results of the Cu/Co multilayer as an example.

The peak intensity in electron units per atom for a sample containing two elements $A$ and $B$ layered alternately $N$ times with a period of $L$ in the small-angle region is expressed as follows:

$$I(Q) = F(Q)F^*(Q)\frac{\sin^2(NQL/2)}{\sin^2(QL/2)}. \tag{8.3}$$

$F$ is the so-called structure factor of the multi-layered sample, and $F^*$ is its complex conjugate. This intensity shows a sharp maximum at the positions of $Q = 2\pi n/L$, where $n$ is an integer and the corresponding structure factor of the $n$-th order peak is provided by the concentration profile, $c(x)$, with depth, $x$:

$$F_n = 2(n_A f_A - n_B f_B)\int_0^{L/2} c(x)\cos\frac{2\pi n x}{L}\mathrm{d}x, \tag{8.4}$$

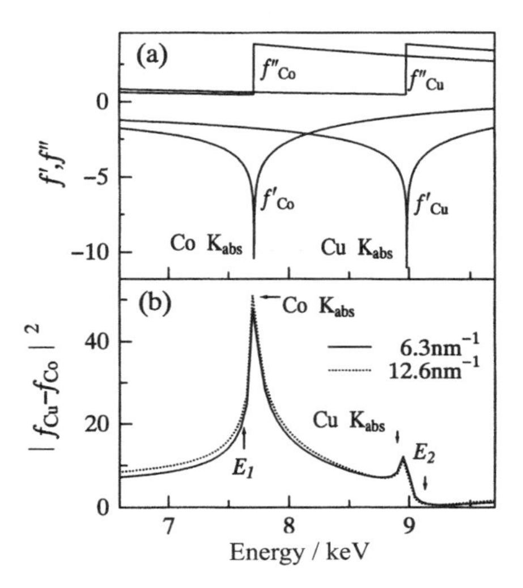


**Fig. 8.7.** **a** Theoretical anomalous dispersion terms of Co and Cu near their K absorption edges. **b** The energy dependence of $|f_{\rm Cu} - f_{\rm Co}|^2$ in the same energy region, where the *solid and dotted lines* correspond to $Q = 6.3$ and 12.6 nm$^{-1}$ and $E_1$ and $E_2$ are the energies selected in the AXS measurements [24]

where $n_j$ and $f_j$ are the number density and the X-ray atomic scattering factor of the $j$-th element, respectively.

The anomalous dispersion term indicates a large variation near it's absorption edge, as shown in Fig. 8.7a using the cases of Cu and Co [24]. Then, the peak intensity, which is roughly proportional to $|f_A - f_B|^2$, shows a distinct change near their absorption edges as illustrated in Fig. 8.7b. It may be worthy of note that the closer the atomic number of each element, the more prominent this change. As seen in Fig. 8.7b, the peak intensity is strongly amplified in the close vicinity of the K absorption edge for Co, and on the other hand the intensity becomes almost equal to zero in the region just above the K absorption edge for Cu. This figure implies that a distinct peak intensity is observed at $E_1$ and almost no peak is observed at $E_2$. In other words, by taking the difference between these intensity profiles, the background intensity can be properly subtracted and the peak profile would be accurately determined without being disturbed by a large background intensity. This idea for the anomalous SAXS method recently proposed by Kato et al. [24] would greatly improve the signal-to-noise ratio especially at a higher-order peak.

Figure 8.8 provides the peak profiles obtained from the intensities of Cu/Co multilayers measured at the two energies of 7.690 and 9.200 keV using the anomalous SAXS method. The peak profiles obtained by the conventional method is also shown in this figure for comparison. The background intensity

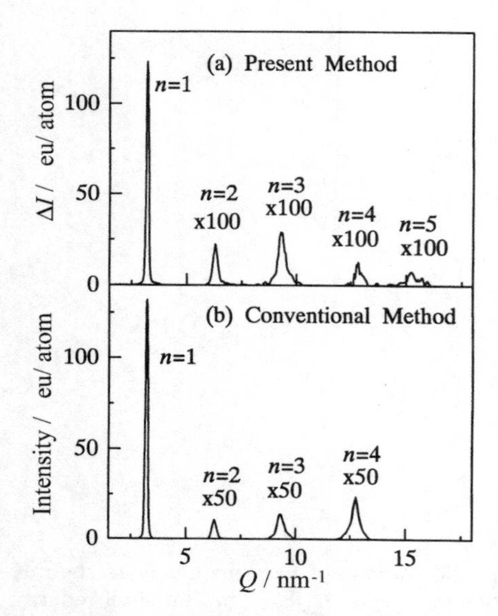


**Fig. 8.8.** The peak profiles of a Cu/Co multilayer obtained from the intensities at 7.690 and 9.200 keV together with those obtained by the conventional method [24]

is much larger than the peak intensity at the higher-order peak. In the conventional method, the lower and upper background intensities are manually determined on either side of each peak. For these reasons, larger experimental errors may be involved during the determination of the background intensities at the higher-order peaks.

As shown in Fig. 8.8, the anomalous SAXS method enables us to detect even a very weak peak of $n = 5$ by making the determination of the background intensity profile precisely. The relative peak intensities of $n = 3$ and $n = 4$ are different in the two methods. The peak intensity for $n = 4$ is about one third of that for $n = 3$ in the anomalous SAXS results (Fig. 8.8a). In contrast, the peak intensity for $n = 4$ is larger than that in the $n = 3$ case in the results obtained by the conventional method (Fig. 8.8b). As explained in Fig. 8.7, the peak intensity is expected to be drastically reduced in the measurement at $E_2$. Kato et al. [24] report that such a change is not observed for the peak of $n = 4$, although the peak intensity for $n = 1$ is extremely reduced and those for $n = 2$, 3 and 5 completely disappear. This variation of the peak intensities at $E_1$ and $E_2$, near the absorption edges, suggests that a large part of the peak intensity for $n = 4$ at about $Q = 13\ \mathrm{nm}^{-1}$ is not attributed to the periodic structure of the Cu/Co multilayer. In other words, an extremely small intensity is detectable by making accurate evaluation of the background possible.

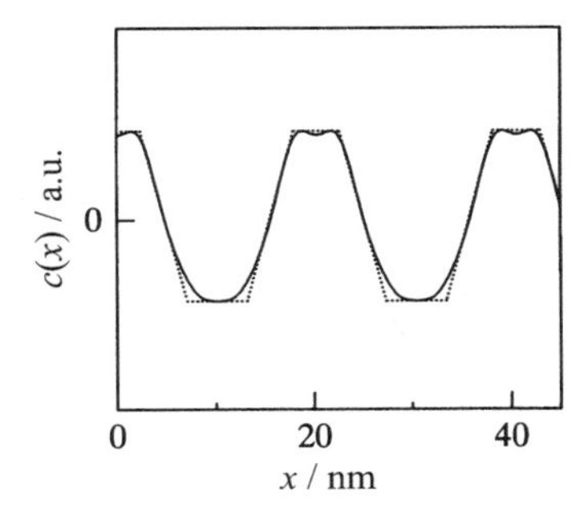


**Fig. 8.9.** The concentration profile in a Cu/Co multilayer obtained by the Fourier transformation of the relative integrated intensities in Fig. 8.8a [24]

Since the integrated intensity is proportional to $FF^*(= |F|^2)$ as shown in (8.3), the concentration profile, $c(x)$, could be obtained by the Fourier transformation of the square root of the differential integrated intensity of Fig. 8.8, as follows:

$$c(x) = \frac{2}{L}\sum_{n=1} \frac{|\Delta F_n|\phi_n}{|\Delta f|} \cos\frac{2\pi n x}{L}, \tag{8.5}$$

$$|\Delta F_n| = \sqrt{|F_n(E_1)| - |F_n(E_2)|}, \tag{8.6}$$

$$|\Delta f| = \sqrt{|n_A f_A(E_1) - n_B f_B(E_1)|^2 - |n_A f_A(E_2) - n_B f_B(E_2)|^2}, \tag{8.7}$$

where $\phi_n$ is the phase factor.

Figure 8.9 shows $c(x)$ for the Cu/Co multilayer calculated from the differential integrated intensity using (8.5), and (8.7). By fitting the experimental $c(x)$ with a trapezoidal profile, it was found that the periodic thickness is 2.045 ± 0.020 nm and that the thicknesses of the Cu and Co layers are 0.455 ± 0.005 nm and 0.545 ± 0.005 nm, respectively. The reliability of these periodic structure parameters of a multilayer is considered to be superior to that of the conventional method. It is not overemphasized because of a considerable reduction in the experimental uncertainties for the background intensity at the higher-order peak and sufficiently reliable detection of the extremely small peak intensity.

It may be helpful to recall some essential points of the new anomalous SAXS method for structural characterization of a multilayered structure. This new method described with the results of the Cu/Co multilayer differs from the conventional AXS case, where the anomalous dispersion effect is used for amplifying the difference of the scattering factors between individual layers so as to enhance the peak intensity. In this method, the peak intensity increases at a certain energy near the absorption edge due to the anomalous dispersion term and at the same time decreases at another energy. Thus, by taking the difference between the intensity profiles measured at these two energies, only the intrinsic peak profiles from a multilayer can be accurately determined. Because of the particular AXS effect, this anomalous SAXS method is the most effective for the multilayered samples consisting of neighboring elements in the periodic table.

Multilayered structures have also been applied as model systems to study inter-diffusion on a very short length and time scale, where the conventional methods become impracticable [25]. In this case, the following equation with respect to the X-ray intensity diffracted at the modulation wave vector, $Q$, is used for measuring the diffusion coefficient, $D$, in multilayered structures:

$$\ln\left(\frac{I(Q)}{I_\circ}\right) = -2DQ^2t, \tag{8.8}$$

where $I_\circ$ and $t$ denote the reference X-ray intensity and time, respectively. The major advantage of this method is that one can measure very low diffusion coefficients. However, this simple analysis relies on the assumption that the two components of the multilayer have the same specific volume. Thus, accurate determination of the specific volumes in multilayers is strongly required. With regard to this subject, Simon et al. [26] devised a technique using the anomalous SAXS. The integrated intensity, $I$, may be proportional to the square of the difference of electron densities:

$$I \propto f_1/V_1 - f_2/V_2, \tag{8.9}$$

where $f_i/V_i$ is the ratio of the atomic scattering factor to the specific volume of the i-th element, which corresponds to the apparent electron density. Therefore, the ratio of the specific volumes can be estimated from a set of intensity profiles measured at energies close to the absorption edge of one constituent.

Figure 8.10 shows the intensity profiles of an Nd–Fe multilayer recorded at energies near the $L_{III}$ absorption edge (6.028 keV) of Nd and the K absorption edge (7.112 keV) of Fe. In both cases, the anomalous dispersion effect is clearly detected, and this anomalous SAXS method was enough to estimate specific volume ratios with an experimental uncertainly of 5% [26]. This method has recently been extended to study the atomic relaxation in the Nd–Fe multilayer during Nd–Fe inter-diffusion [27].

## 8.4 Morphology of a Sulfonated Polystyrene Ionomer

Ionomers consisting of hydrocarbon polymers containing less than about 10 mol.-% of acid-salt side groups have recently stirred technological and scientific interest due to their unique properties arising mainly from electrostatic or dipole interactions of the ionic species [28, 29, 30]. Most ionomers are considered micro-phase-separated aggregates rich in some ionic species formed by

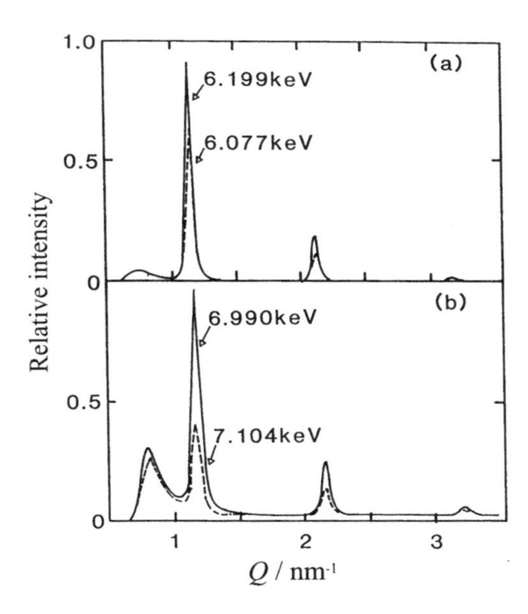


**Fig. 8.10. a, b** Rocking curves of an Nd-Fe multilayer ($d_{Nd}$ = 3.7nm, $d_{Fe}$ = 1.6nm). **a** Recorded near the $L_{III}$ absorption edge of Nd and **b** recorded near the K absorption edge of Fe (Note the flat background due to Nd fluorescence) [26]

the strong interactions between the ionic monomer units. In order to characterize such structural features, the SAXS method is considered to be a useful technique. However, information about the morphology of the ionomer is required before the relationship between the SAXS intensity data and physical, chemical and transport properties can be assessed.

Typical SAXS intensity patterns from ionomers are known to indicate a peak with a Bragg space usually ranging from 2 to 10 nm and a strong upturn of scattering intensity close to the angular origin. Although several models for morphology of ionomers have been proposed, none of them can fit both the peak and the upturn with a consistent set of parameters. This inconvenience is attributed to the reason that the interpretation of any SAXS intensity data heavily depends on the assumption used in the model. The use of the anomalous dispersion effect enables us to provide one way to reduce such difficulty by changing the scattering power of the specific element selectively and thus separating the scattering contribution involving cations from non-ionic scattering sources. In other words, one can obtain information "whether the cation in the ionomer gives rise to the observed upturn as well as the peak in the SAXS data".

Figure 8.11a shows the anomalous SAXS results for a nickel-neutralized sulfonated polystyrene ionomer measured at energies of 5 and 100 eV below

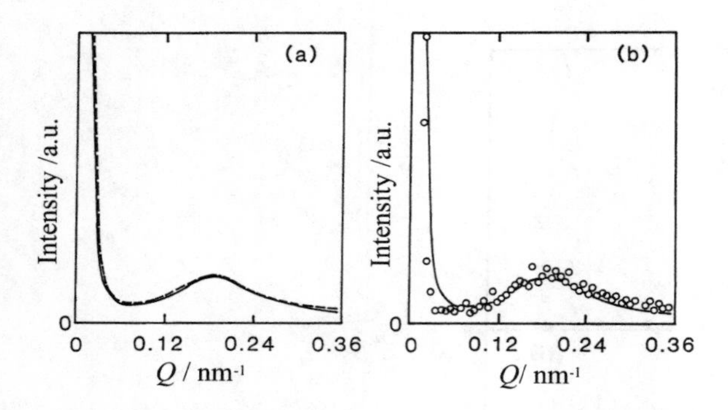


**Fig. 8.11. a, b** SAXS intensity data for nickel-neutralized sulphonated polystyrene. **a** SAXS intensities measured at 100 (*solid line*) and 5 eV(*dashed line*) below the K absorption edge of Ni. **b** SAXS intensity data measured at 100 eV(*solid line*) below the edge; *circles* denote the difference pattern in **a**. The difference pattern has been scaled to equal the solid curve in the integrated intensity [31]

the K absorption edge (8.333 keV) of Ni, and Fig. 8.11b provides the difference pattern (circles) together with the SAXS intensity data at 100 eV below the edge of Ni (solid line) [31]. The anomalous dispersion effect is 8% of the total intensity for the nickel-neutralized ionomer, as compared with 23% for the pure Ni metal [31], although the observed magnitude of the anomalous dispersion effect might be affected, more or less, by the ionic aggregates consisting of not pure $Ni^{2+}$ cations. It is also expected to contain sulfonate groups, which have a higher electron density than the polystyrene matrix.

The upturn of scattering intensity close to the angular origin is still present, suggesting that it must be related to the neutralizing cations, because if it is attributed to a source devoid of cations, it would have cancelled out when calculating the difference pattern. This may be caused by an inhomogeneous distribution of isolated ionic groups in the nickel-neutralized sulfonated polystyrene ionomer. It is worth mentioning that the ionomer peak is also observed in the difference pattern and appears similar to that in the original SAXS curve. This simply suggests that the ionomer peak reflects the scattering from the ionic aggregates containing nickel cations.

Metal-supported catalysts are usually three-phase systems consisting of metal particles, the substrate support and its pores. For example, they are prepared from electronically conductive porous carbon with noble metal catalysts such as Pt and Pd–Au alloys in the form of small crystallites with diameters of around 1 to 5 nm and corresponding atomic fractions of about $10^{-3}$. However, the conventional SAXS method cannot separate the scattering contribution of metal particles from the background scattering of the topological pore structure; therefore, it is difficult to obtain information about the number of active sites and the size distribution of particles. This experimental inconvenience can be overcome by the anomalous dispersion effect, where the

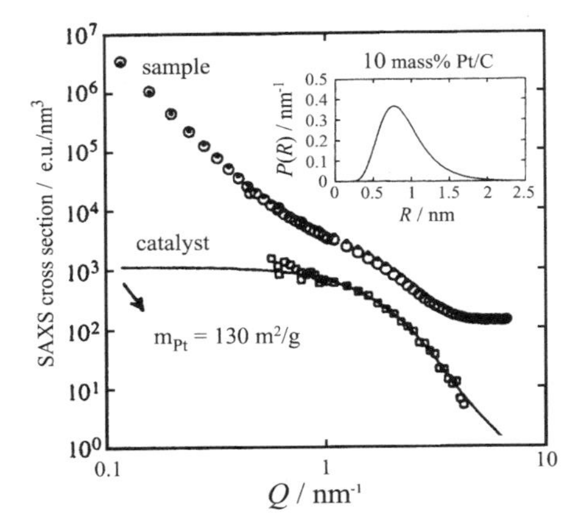


**Fig. 8.12.** Anomalous SAXS results of the carbon-supported 10 mass-% Pt catalyst measured at energies of 10.353 and 11.548 keV near the $L_{III}$ absorption edge (11.563 keV) of Pt [33]. The *solid line* is a fit of the log-normal size distribution assuming the mean radius $R_\circ = 0.86$ nm and mean particle volume 3.7 nm$^3$. *Inset*: resultant log-normal size distribution of the Pt catalyst particles

scattering factor of metal varies at its absorption edge, whereas the scattering factors of the non-metal substrate components remain constant [32, 33]. Therefore, the contribution of the catalyst particles can be straightforwardly separated from the measured intensity contrast.

Haubold et al. [33] reported anomalous SAXS results of carbon-supported 10 mass-% Pt catalyst measured at the $L_{III}$ absorption edge (11.563 keV) of Pt. The following two remarks were presented and the size distribution and particle volume were estimated: In the smaller wave vector region, mainly attributed to the contribution from pores in the carbon support, no energy dependence was detected. On the other hand, scattering contrast was obtained in the larger wave vector region. This arises from the catalyst scattering of nanometer-sized Pt particles. The results of Haubold et al. [33] are summarized in Fig. 8.12. The solid line in this figure is a fit of the log-normal size distribution of the mean radius $R_\circ = 0.86$ nm and mean particle volume 3.7 nm$^3$. The results clearly demonstrate the usefulness of the anomalous SAXS method for characterizing individual sizes and distributions of nanometer-sized metal catalysts formed on a porous support by obtaining the scattering contrast with sufficient reliability.

## References


1. O. Glatter and O. Kratky: *Small-Angle X-Ray Scattering* (Academic Press, New York 1982)
2. P. Goudeau, A. Fontaine, A. Naudon and C.E. Williams: J.Appl. Cryst., **19**, 19 (1986)
3. O. Lyon, J.J. Hoyt, R. Pro, B.E.C. Davis, B. Clark, D. de Fontaine, and J.P.Simon: J.Appl.Cryst.,**18**, 480 (1985)
4. O. Lyon, and J.P. Simon: Acta Metal., **34**, 1197 (1986)
5. N. Bouzid, C. Servant and O. Lyon: Philos Mag., B **57**, 343 (1988)
6. C. Servant and N. Bouzid: Acta Metall. **36**, 2771 (1988)
7. S. Djanarthany, C. Servant and O. Lyon: Philos. Mag. A **66**, 575 (1992)
8. F.A. Sadi and C. Servant: Philos. Mag., A **80**, 639 (2000)
9. M. Rice, S. Wakatsuki and A. Bienenstock: J. Appl. Cryst., **24**, 598 (1991)
10. M.J. Regan and A. Bienenstock: Phys. Rev., **51**, 12170 (1995)
11. T. Zhang, A. Inoue, S. Chen and T. Masumoto: Mater Trans., JIM **33**, 143 (1993)
12. K. Sugiyama, A.H. Shinohara, Y. Waseda, S. Chen and A. Inoue: Mater Trans. JIM., **35**, 481 (1994)
13. P. Goudeau, A. Naudon, A. Chamberod, B. Rodmacq and C.E. Williams: Europhys. Lett., **3**, 269 (1987)
14. E. Matsubara, K. Sugiyama, A.H. Shinohara, Y. Waseda, A. Inoue, T. Zhang and T. Masumoto: Mater. Sci. Eng., A **179/180**, 444 (1994)
15. K. Sugiyama, A.H. Shinohara, Y. Waseda and A. Inoue: J. Non-Cryst. Solids, **192/193**, 376 (1995)
16. M. Maret, J.P. Simon, B. Boucher, R. Tourbot and O. Lyon: J. Phys.: Condens. Matter, **4**, 9709 (1992)
17. M. Maret, J.P. Simon and O. Lyon: J.Phys: Condens Matter, **1**, 10249 (1989)
18. D.H. Mosca, F. Petroff, A. Fert, P.A. Schreder, W.P. Pratt Jr., R. Laloee and S. Lequien: J.Magn. Mater., **94**, L1 (1991)
19. S.S. Parkin, R. Bhadra and K.P. Roche: Phys. Rev. Lett., **66**, 2152 (1991)
20. H.E. Fischer, H. Fischer, G. Durand, G. Pellegrino, S. Andrieu, M. Piecuch, S. Lefebvre, J. Bessière: Nuclear Instrum. Meth. Phys. Res. B **97**, 402 (1995)
21. J.M. Mariot, J.J. Gallet, L. Journel, C.F. Hague, W. Felsch, G. Krill, M. Sacchi, J.P. Kappler, A. Rogalev, J. Goulon: Physica B **259–261**, 1136 (1999)
22. K. Ohsumi: J. Crystallogr. Soc. Jpn., **27**, 73 (1985)
23. N. Nakayama, I. Moritani, T. Shinjo, Y. Fujii and S. Sasaki: J.Phys. F: Met. Phys. **18**, 429 (1988)
24. K. Kato, E. Matsubara, M. Saito, T. Kosaka, Y. Waseda and K. Inomata: Mater. Trans. JIM, **36**, 408 (1995)
25. A. Bruson, M. Piecuch and G. Marchal: J. Appl. Phys., **58**, 1229 (1985)
26. J.P. Simon, O. Lyon and M. Piecuch: J. Appl. Cryst., **21**, 317 (1988)
27. J.P. Simon, O. Lyon and A. Bruson: J. Appl. Cryst., **24**, 156 (1991)
28. R.A. Weiss and G.A. Lefelar: Polymer, **27**, 3 (1986)
29. W.J. Macknight and T.R. Earnest, Jr.: J. Polymer Sci., Macromol. Rev, **16**, 41 (1981)
30. C.G. Bazuin and A. Eisenberg: Ind. Eng. Prod. Res. Dev., **20**, 271 (1981)
31. Y.S. Ding, S.R. Hubbard, K.O. Hodgson, R.A. Register and St.L. Copper: Macromolecular, **21**, 1698 (1988)
32. H-G. Haubold and X.H. Wang: Nucl. Instrum. Meth. Phys. Rev., B **97**, 50 (1995)
33. H-G. Haubold, X.H. Wang, G. Goerigk and W. Schilling: J. Appl. Cryst., **30**, 653 (1997)

# 9. Anomalous Grazing-Incidence X-ray Reflection

The production of multi-layered thin films such as those of Co/Cu with sufficient reliability is well recognized as a key technology for device fabrication in micro-electronics, because of their characteristic gigantic magnetoresistance and antiferro- and ferro-magnetic oscillation behaviors as a function of the thickness of the layers of the non-magnetic component [1, 2]. These interesting properties of new synthetic functional materials should be attributed to their periodic and interfacial structures at a microscopic level, although the origin of such peculiar feature of multi-layers is not well identified yet. This includes knowledge of the surface structure and the number density of atoms in the near-surface region of the desired materials. On the other hand, knowledge of the structural properties of surfaces and interfaces of materials is a prerequisite to countless applications originating from physics, chemistry and engineering. This demand has been satisfied by the availability of a large variety of surface-specific experimental methods. For example, electron diffraction techniques such as RHEED are often used for the surface analysis because of their inherent surface sensitivity due to the strong interaction between electrons and atoms. However, the limitation of these techniques is that some tedious experimental conditions, such as ultra-high vacuum and elaborate sample preparations, are required.

The use of X-ray scattering or X-ray reflection under the condition of grazing incidence has experienced a massive upsurge which has been made possible by the development of a powerful X-ray source of synchrotron radiation. Two typical diffraction geometries of X-rays for a surface are given in Fig. 9.1. The one geometry in Fig. 9.1a is the Seemann–Bohlin [3] type, widely used for structural analyses of thin films grown on a substrate. Another geometry in combination with the total external reflection is the grazing-incidence X-ray scattering [4] (hereafter referred to as GIXS) type illustrated in Fig. 9.1b, in which information on a depth profile of a surface structure is obtained by making effective use of a drastic change in penetration depth across the critical angle [5,6,7]. In order to facilitate the understanding of these experimental conditions, a schematic of all angular motions of a sample and detectors in the GIXS-type X-ray spectrometer is illustrated in Fig. 9.2 [8]. This GIXS-type spectrometer consists of two double-axis diffractometers. One double-axis diffractometer, labeled A in Fig. 9.2, is horizontally placed and has two

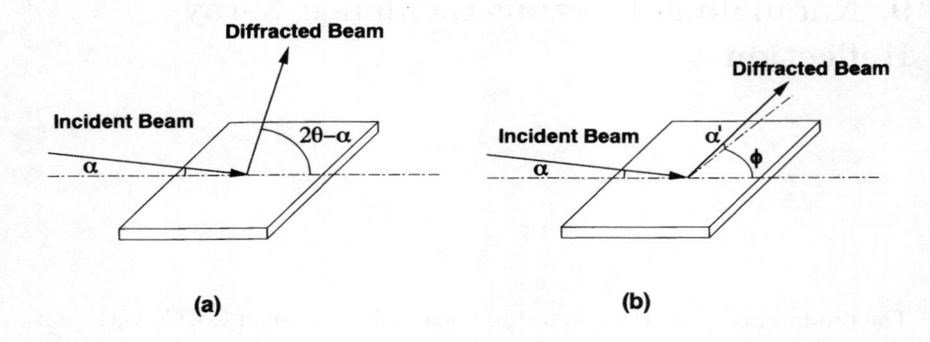


**Fig. 9.1. a, b** Schematic of typical diffraction geometry of X-rays for a sample surface: **a** The Seemann–Bohlin geometry and **b** the GIXS geometry

rotation axes, $\alpha$ and $2\theta$. Another diffractometer, labeled B, with two rotation axes, $\omega$ and $\phi$, is vertically placed on the $\alpha$ rotation axis of diffractometer A. The position of each diffractometer is precisely set by adjusting tools of $xy$ translations as well as a height control table. In this way, the incident beam passes through the intersection of the $\alpha$ and $\omega$ rotation axes. A film sample is mounted on a holder in diffractometer B equipped with a crossed tilting stage ($\delta'$ and $\delta''$) and a vertical translation, so that the surface normal of the sample is aligned with the $\omega$-axis. The sample surface is also adjusted to coincide with a plane including the incident beam and the $\alpha$-axis. An incident angle, $\alpha$, can be determined by the $\alpha$ rotation of diffractometer A.

It may be worth mentioning that the measurement with the Seemann–Bohlin geometry (see, Fig. 9.1a) is carried out using a fixed incident angle, $\alpha$, set by rotation around the $\alpha$-axis and a scan of the counter A around the $2\theta$-axis. The other geometry, drawn in Fig. 9.1b, is the so-called GIXS geometry, where $\alpha$ is similarly selected by $\alpha$ rotation, and the counter B is scanned around the $\phi$-axis with a proper glancing angle, $\alpha'$. Under these conditions X-rays provide a non-destructive probe for analyzing the surface structure in a range from a few nm to a hundred nm with depth controlled sensitively. Another merit of this technique is that the weak interaction of X-rays with condensed matter allows us to interpret the experimental data easily within the framework of kinematic theory of diffraction. For these reasons, X-ray optical methods such as grazing-incidence X-ray diffraction (GIXD) and grazing X-ray reflectometry (GXR) are now widely used to investigate the surface structure of various materials, including multi-layered thin films. It should be stressed, however, that these methods often require with the highest priority the determination of the atomic number density of the constituents, especially in unknown materials, in order to prepare an essential condition as input for structural characterization. X-ray photoelectron spectroscopy (XPS), Auger electron spectroscopy (AES) and secondary ion mass

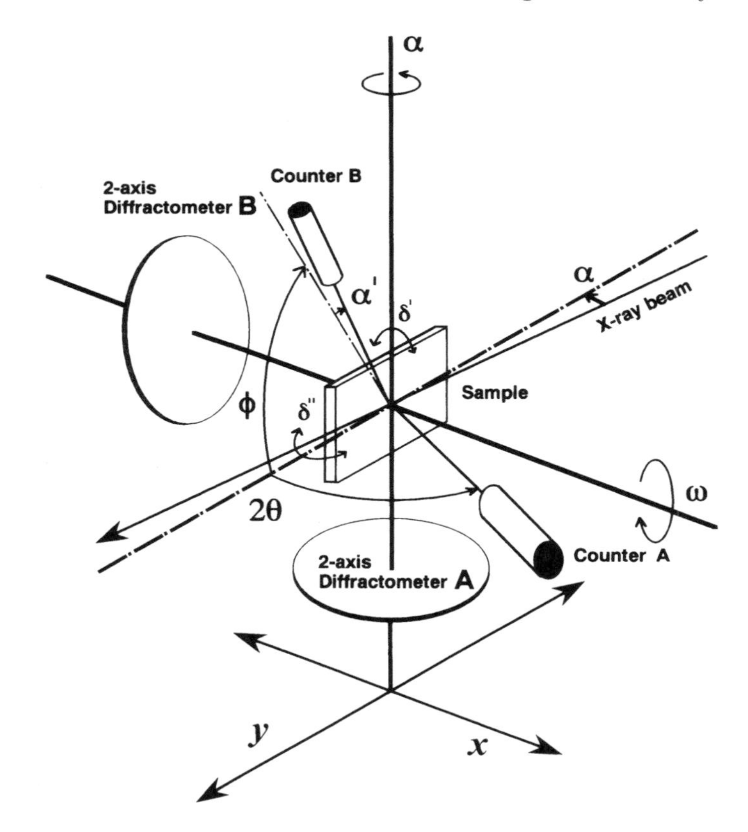


**Fig. 9.2.** Schematic of configurations of rotation axes used in the crossed double-axis diffractometer

spectroscopy (SIMS) are extensively applied for determining the composition of thin films with high sensitivity to the surface, but they give only relative quantities of constituents. Furthermore, they are destructive probes for obtaining the compositional depth profiles in the sample surface by sputtering.

In this important subject, the anomalous grazing X-ray reflection (hereafter referred to as AGXR) method [9, 10] appears to be one way to determine the absolute value of the atomic number density in materials nondestructively. The AGXR method is based on the idea of measuring the deviation in the refractive index of a substance of interest through the anomalous dispersion phenomena. The usefulness and essential points of the AGXR method are described with some selected examples, including a multi-layered thin film consisting of heterostructure GaAs/AlAs/GaAs.

## 9.1 Fundamentals of the AGXR Method

It is well known that the refractive index for X-rays of energy $E$ is given by the following equation [11]:

$$n(E) = 1 - \delta(E) + \mathrm{i}\beta(E), \tag{9.1}$$

in which

$$\delta(E) = \frac{\lambda^2 r_\mathrm{e}}{2\pi v_\mathrm{c}} \sum_j [Z_j + f_j'(E)], \tag{9.2}$$

$$\beta(E) = \frac{\lambda^2 r_\mathrm{e}}{2\pi v_\mathrm{c}} \sum_j [-f_j''(E)], \tag{9.3}$$

where $\lambda$ is the X-ray wavelength, $r_\mathrm{e}$ is the classical electron radius, $v_\mathrm{c}$ is the volume of unit cell, $Z_j$ is the atomic number, $f_j'(E)$ and $f_j''(E)$ are the real and imaginary parts of the anomalous dispersion term and $\rho_j$ is the atomic number density of the $j$-th element. The value of $\delta(E)$ is positive in any case, so $n(E)$ is less than unity. Thus, total external reflection occurs for angles of incidence, $\alpha$, lower than the critical angle, $\alpha_\mathrm{c}(E)$. When the absorption for X-rays is quite small, the critical angle is simply given by

$$\alpha_\mathrm{c}(E) = \sqrt{2\delta} = \lambda\sqrt{\frac{r_\mathrm{e}}{\pi v_\mathrm{c}} \sum_j [Z_j + f_j'(E)]}. \tag{9.4}$$

Since $[Z_j + f_j'(E)]/M_j$ (where $M_j$ is an atomic weight of the $j$-th atom) is approximately equal to 1/2, the critical angle is directly related to the density, $\rho$, by the following equation:

$$\alpha_\mathrm{c}(E) \approx \lambda\sqrt{\frac{r_\mathrm{e}\mathrm{N_A}}{2\pi}\rho}, \tag{9.5}$$

where $\mathrm{N_A}$ is Avogadro's number. Incident X-rays are known to be evanescent within the substance and penetrate only a few nm below $\alpha_\mathrm{c}$. The $\mathrm{e}^{-1}$ penetration depth, $D(\alpha)$, depends upon $\alpha$ and is given by [4, 12]

$$D(\alpha) = \frac{\lambda}{4\pi q}, \tag{9.6}$$

where

$$q = \frac{1}{\sqrt{2}}\sqrt{\sqrt{(\alpha^2 - \alpha_\mathrm{c}^2)^2 + 4\beta^2} + \alpha_\mathrm{c}^2 - \alpha^2}. \tag{9.7}$$

The X-ray penetration depth changes from a few nm to several hundred nm by varying the incident angle from below to above $\alpha_\mathrm{c}$, which enables us to obtain the depth profile of a surface structure. $D(\alpha)$ is given in Fig. 9.3, using the results of austenite stainless steel for Mo-K$\alpha$ radiation It is easily found that $D(\alpha)$ sharply varies across $\alpha_\mathrm{c} = 0.18°$.

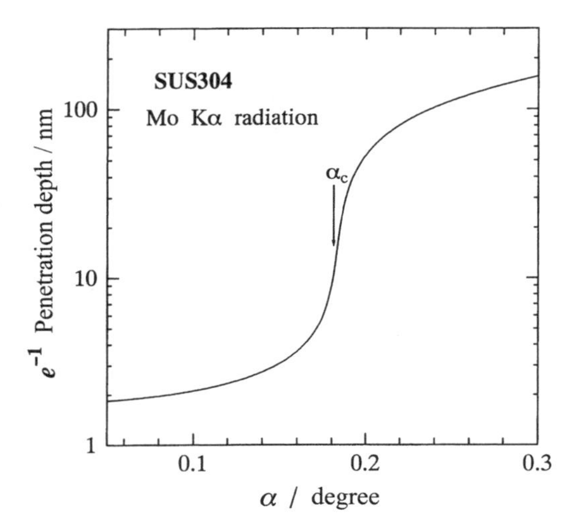


**Fig. 9.3.** Penetration depth in the austenite stainless steel for Mo-K$\alpha$ radiation

The AGXR method is based on the idea of measuring the deviation in the refractive index of a substance of interest through the anomalous dispersion phenomena. The schematic for the experimental mode given in Fig. 9.4 may facilitate the understanding of the AGXR concept for determining the number density of atoms in the near-surface region.

When the summation over all atoms in the unit cell in (9.4) is converted to the sum over all elements, the critical angle can be rewritten and related to the anomalous dispersion term in the following form:

$$\frac{\alpha_c(E)}{\lambda} = \sqrt{\frac{r_e}{\pi} \sum_k \rho_k [Z_k + f'_k(E)]}. \tag{9.8}$$

As the incident X-ray energy approaches the absorption edge, $\alpha_c$ should be shifted to a lower angle because the effective number of electrons closely connected to the scattering of X-rays is significantly reduced due to the anomalous dispersion effect. For this reason, a magnitude of its angular deviation heavily depends upon the number of resonating atoms in a substance. Let us consider the measurement of critical angles at two different energies, $E_1$ and $E_2$, close to the absorption edge, $E_{abs}$, of one of the constituents, for example, the element $A$. The energy variation of the scattering factors including anomalous dispersion terms for constituent elements except for $A$ is known to be negligibly small; the following simple and useful relation is readily obtained [9]:

$$\Delta \left( \frac{\alpha_c(E)}{\lambda} \right)^2 \equiv \left( \frac{\alpha_c(E_1)}{\lambda_1} \right)^2 - \left( \frac{\alpha_c(E_2)}{\lambda_2} \right)^2 \equiv k_c(E) \tag{9.9}$$

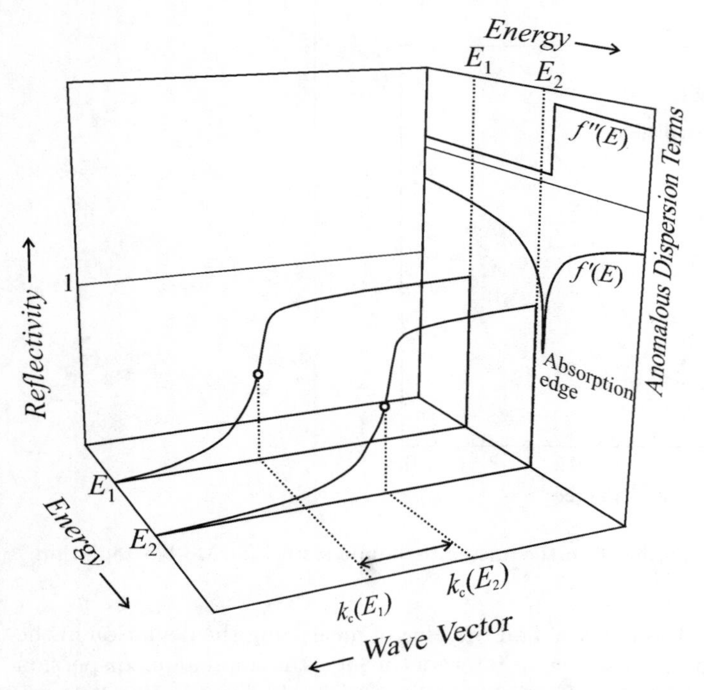


**Fig. 9.4.** Schematic of the experimental mode by the AGXR method

$$\cong \rho_A \frac{r_e}{\pi}[f'_A(E_1) - f'_A(E_2)]. \tag{9.10}$$

When plotting the relation between $[\Delta(\alpha_c(E)/\lambda)^2] = k_c(E)$ and $[f'_A(E_1) - f'_A(E_2)]$, a straight line is predicted, and its slope directly gives the atomic number density, $\rho_A$, of $A$. The penetration depth of X-rays into the materials at the critical angle, expressed by $D(\alpha_c) = \lambda/4\pi\beta^{1/2}$, is considered to be of the order of a few tens of nm. Therefore, it should be kept in mind that the number density value estimated by the AGXR method appears to give an average in depth ranging to a few tens of nm at most. This AGXR procedure has the advantage that it is not necessary to pay full attention to the determination of the absolute angles; because the experimental uncertainty in angle is canceled out through the process for calculating the value of $\Delta(\alpha_c(E)/\lambda)^2$.

## 9.2 Zr and Y Atoms in the Surface of a Sintered $ZrO_2$–$Y_2O_3$ Crystal Plate

The surface of a sintered $ZrO_2$–3 mol.-%$Y_2O_3$ plate of single tetragonal phase with $a = 0.5095$ nm and $c = 0.5177$ nm (TEP Co., Ltd) was mechanically polished with a diamond paste until it showed plainly total external reflection of

X-rays. The reflection measurements using the experimental setup described in Chap. 5 were carried out with synchrotron radiation at a 6B beam line in the Photon Factory of the High Energy Accelerator Research Organization, Tsukuba, Japan.

Figure 9.5 shows reflection intensity profiles of $ZrO_2$–3 mol.-%$Y_2O_3$ as a function of $(\alpha/\lambda)^2$ measured at incident energies of –25, –50, –80, –150 and –300 eV away from the Zr-K (17.999 keV) and Y-K absorption edges (17.040 keV). The shifts of critical angle are well recognized in the profiles measured at the Zr-K absorption edge, whereas such variation is not clearly detected in the profiles measured at the Y-K absorption edge. The difference in magnitude of their shifts can be attributed to the content of Zr or Y. Then, the values of $\alpha_c(E)$ in each curve are chosen so as to give the best fit between the calculated profile and experimental data [13]. The values of $(\alpha_c(E)/\lambda)^2$ determined by the nonlinear least-squares fitting are plotted as a function of $f'(E)$ in Fig. 9.6, and we find a linear relationship. The atomic number densities of Zr and Y in the surface of $ZrO_2$–3 mol.-%$Y_2O_3$

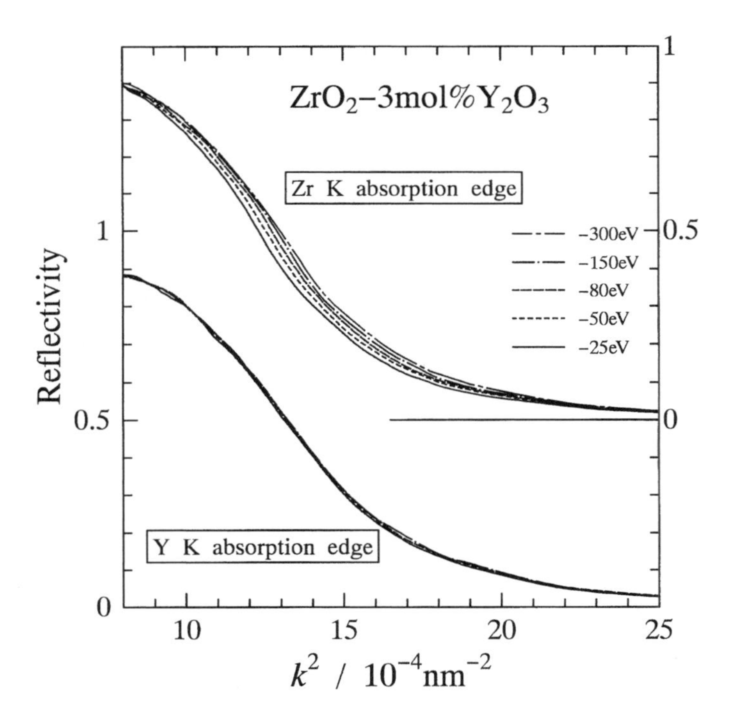


**Fig. 9.5.** Experimental reflection curves of $ZrO_2$–3 mol.-%$Y_2O_3$ measured at the lower-energy side of the Zr-K (17.999 keV) and Y-K absorption edges (17.040 keV) [9]

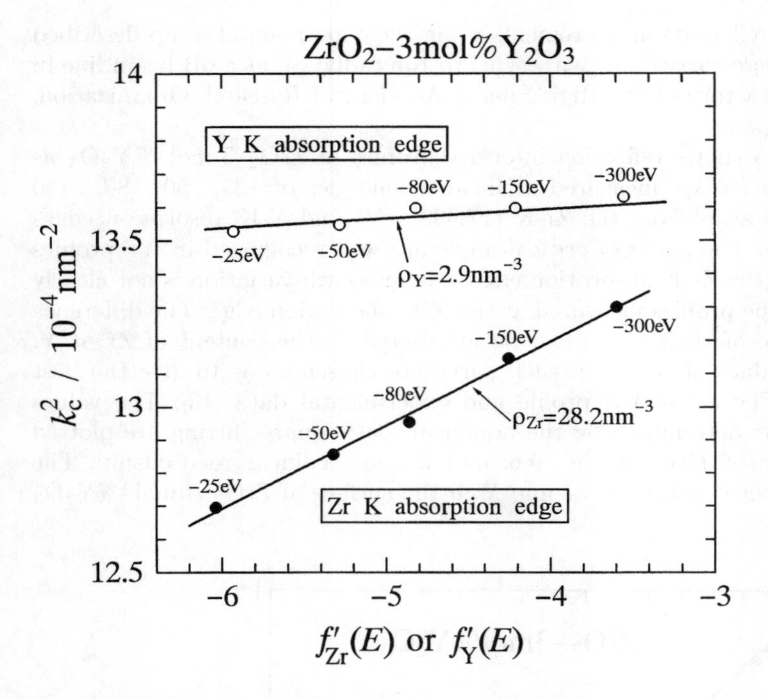


**Fig. 9.6.** Plots of $(\alpha_c(E)/\lambda)^2$ as a function of $f'_{\text{Zr}}(E)$ or $f'_{\text{Y}}(E)$ [9]

crystal plate were estimated from each slope of the straight lines to be 28.2 $\pm$ 0.7 and 2.9 $\pm$ 1.5 atoms nm$^{-3}$, respectively. On the other hand, the atomic number densities of $\rho_{\text{Zr}} = 28.0$ atoms nm$^{-3}$ and $\rho_{\text{Y}} = 1.8$ atoms nm$^{-3}$ are suggested from calculation using the lattice constants of tetragonal $ZrO_2$–3 mol.-%$Y_2O_3$. The agreement for the Y case appears not to be sufficient. However, the author maintains the view that the atomic number densities in the surface of the $ZrO_2$–3 mol.-%$Y_2O_3$ crystal plate determined by the AGXR method coincide with those of the bulk within the experimental errors.

## 9.3 Cr and Fe Atoms in the Passive Film of Stainless Steel

The AGXR method has been applied to the gold-colored stainless steel named "Lumina Color" (Kawasaki Steel Corp.) [14]. The latter is colored by forming a passive film of chromium–iron oxide on the surface of stainless steel using the alternating current electrolyzing method; the thickness of the oxide layer is 190 nm [8]. This thin oxide layer is considered to be in a non-crystalline state because no sharp Bragg peak was found using the grazing-incidence

X-ray scattering (GIXS) method. The reflection curves of the gold-colored stainless steel are given in Figs. 9.7a, b, measured at the lower-energy side of the Cr-K and Fe-K (7.113 keV) absorption edges, respectively. Each reflection curve gives an oscillating structure which is the result of interference between X-rays reflected from the surface of oxide film and those reflected from the oxide/metal interface inside the film (see Fig. 9.8). Appreciable shifts are detected in the phase of the oscillating parts at both energy regions (see Fig. 9.7). When the reflectivity is small in the higher-angle region above the critical angle, the reflectivity, denoted by $R(\alpha)$, for a single-layer thin film may be expressed by the combination of the non-oscillating and oscillating parts as follows [15, 16]:

$$
\begin{aligned}
R(\alpha) = & \frac{F_{1,2}^2 + F_{2,3}^2}{(1 - F_{1,2}^2)(1 - F_{2,3}^2)} \\
& + \frac{2F_{1,2}^2 F_{2,3}^2}{(1 - F_{1,2}^2)(1 - F_{2,3}^2)} \cos\left(4\pi t \sqrt{(\alpha^2 - \alpha_c^2)/\lambda^2}\right) ,
\end{aligned}
\tag{9.11}
$$

where $F_{1,2}$ and $F_{2,3}$ are the so-called Fresnel coefficients at the air/Cr and Cr/glass interfaces, and $t$ is the thickness of Cr thin layer. According to (9.11), the oscillating part is described by a simple cosine function. Therefore, the phase shift in the oscillating parts of the reflection profiles measured at different energies are rigorously equal to the changes in the critical angles.

The systematic shifts in the phase of the oscillating parts are well appreciated, as seen in Fig. 9.7. After subtracting the non-oscillating part from each reflection profile, the phase shift of the oscillating part of $\Delta(\alpha_c(E)/\lambda)^2$ can be estimated; the values of $\Delta(\alpha_c(E)/\lambda)^2$ are plotted as a function of $f'(E)$ of Cr (open circles) and Fe (closed squares) in Fig. 9.9. Here, the value at the energy corresponding to 25 eV below the absorption edge of $E_{abs}$ was used as a reference in calculating $\Delta(\alpha_c(E)/\lambda)^2$ for convenience. A linear relationship between $\Delta(\alpha_c(E)/\lambda)^2$ and $f'(E)$ is again clearly obtained, and these two slopes provide the values of $\rho_{Cr}$ and $\rho_{Fe}$ in the passive film formed on the gold-colored stainless steel: $\rho_{Cr} = 11.8 \pm 0.4$ and $\rho_{Fe} = 7.4 \pm 0.3$ atoms nm$^{-3}$, respectively. Thus, the concentration ratio of Cr/Fe in the gold-colored stainless steel sample is found to be 1.59. It may be worth mentioning from the micro AES results [14] that the film on the gold-colored stainless steel surface has uniform Cr dispersion and the concentration ratio of Cr/Fe is 1.5 (this value is consistent with the AGXR result).

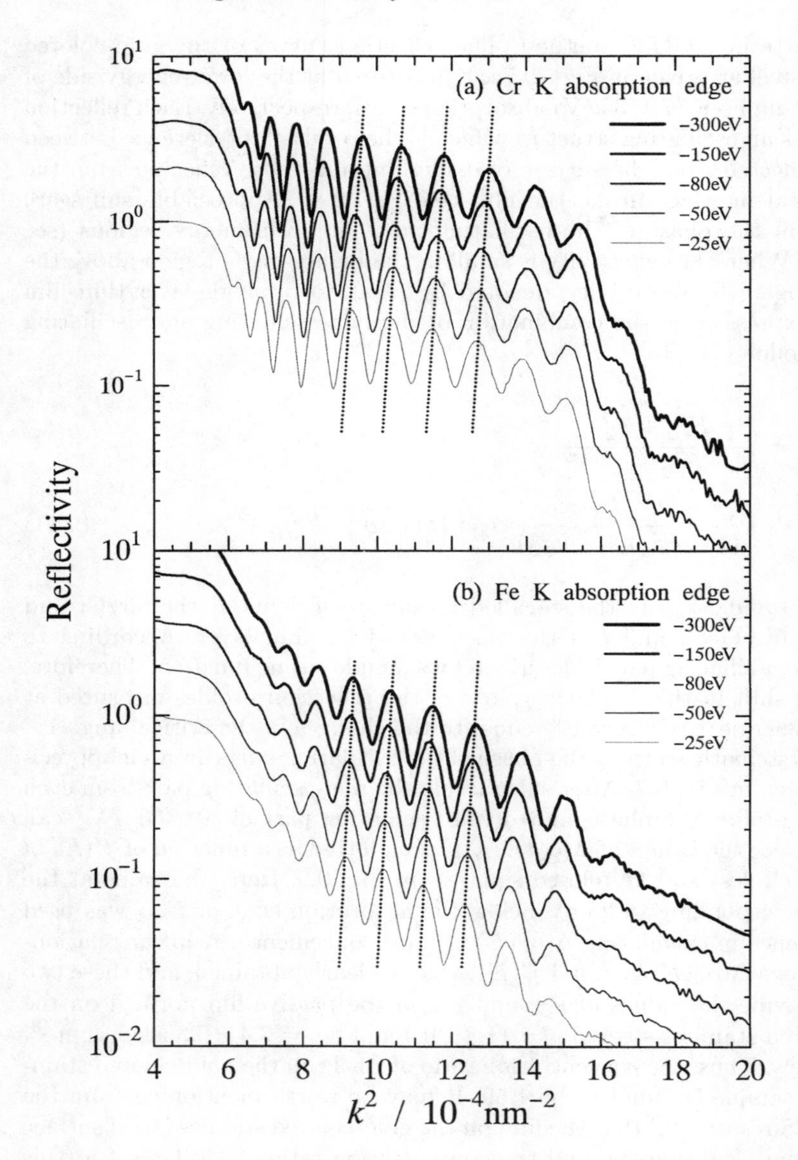


**Fig. 9.7.** Experimental reflection curves of gold-colored stainless steel measured at the lower-energy side of the Cr-K (5.989 keV) and Fe-K absorption edges (7.113 keV) [9]

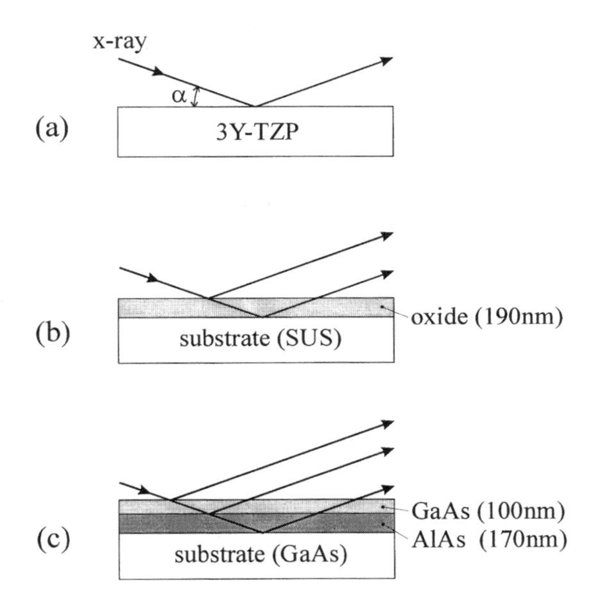


**Fig. 9.8. a-c** Schematic of the experimental conditions for X-ray reflectometry. **a** Bulk surface, **b** oxide thin film grown on a substrate and **c** multi-layered thin film grown on a substrate

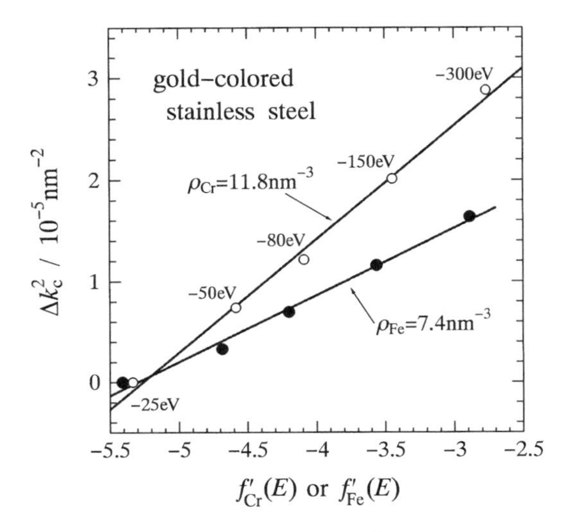


**Fig. 9.9.** Plots of $\Delta(\alpha_c(E)/\lambda)^2$ for gold-colored stainless steel as a function of $f'_{Cr}(E)$ and $f'_{Fe}(E)$ [9]

## 9.4 Ga and As Atomic Number Density in Multi-layered Thin Films

The range of application of the AGXR method is not limited in practice to the bulk surface or a single-layered thin film grown on a substrate. When coupled with the Fourier filtering technique, the AGXR method can be extended to multi-layered thin films. Of course, the principle of the AGXR method itself is unchanged in this extension. The usefulness of this method is demonstrated by analyzing a multi-layered thin film consisting of the heterostructure GaAs/AlAs/GaAs [10].

The sample used here is a GaAs/AlAs heterostructure grown on a (001) GaAs substrate by molecular-beam epitaxy. The sample structure consists of air/GaAs($t_1$ ~100 nm)/AlAs($t_2$ ~170 nm)/GaAs(0.6-mm substrate) as illustrated in Fig. 9.8c. The sample is mounted on a holder, which is attached to a double-axis goniometer and equipped with a crossed tilting stage and a vertical translation, so that the sample surface is adjusted to coincide with a plane including the incident beam. It may be noticed that the finest glancing angle of the incident X-rays is 0.0002° in each step. Figure 9.10 shows the reflection intensity profiles of GaAs/AlAs/GaAs measured at incident energies of –25, –50, –80, –150 and –300 eV away from the Ga-K (10.368 keV) or As-K absorption edges (11.865 keV). Each reflection curve clearly shows an oscillating structure, which is the result of interference between X-rays reflected from each layer interface.

For simplification, let us introduce the wave vector $k$ ($\equiv \alpha/\lambda; \alpha$ is the glancing angle and $\lambda$ is the wavelength). For a double-layer thin film on a substrate, the reflectivity of (9.11) may be re-written in the following form, which includes the energy dependence due to the anomalous dispersion effect [15, 16]; here, the reflectivity, $R(k, E)$, is separated into two parts, a non-oscillating part, $R^\circ(k, E)$, and an oscillating part, $\chi(k, E)$:

$$R(k,E) = R^\circ(k,E) + \chi(k,E), \tag{9.12}$$

$$R^\circ(k,E) = \frac{\left(F_{0,1}^2 + F_{1,2}^2 + F_{2,3}^2 + F_{0,1}^2 F_{1,2}^2 F_{2,3}^2\right)}{\left(1 - F_{0,1}^2\right)\left(1 - F_{1,2}^2\right)\left(1 - F_{2,3}^2\right)}, \tag{9.13}$$

$$\begin{aligned} \chi(k,E) &= \sum_i \chi_i(k,E) \\ &= A\cos\gamma_1 + B\cos\gamma_2 + C\cos(\gamma_1+\gamma_2) \\ &+ D\cos(\gamma_1-\gamma_2) + \cdots, \end{aligned} \tag{9.14}$$

$$\gamma_j = 4\pi t_j \sqrt{k^2 - k_{cj}^2(E)}, \tag{9.15}$$

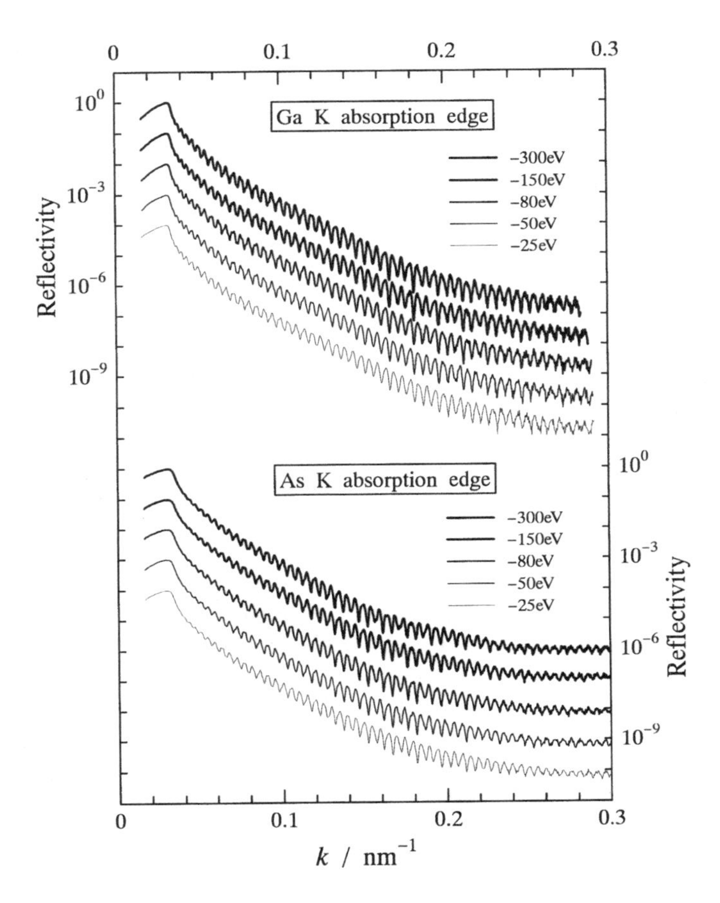


**Fig. 9.10.** Experimental reflection curves of GaAs/AlAs/GaAs measured at the lower-energy side of the Ga-K (10.368 keV) and As-K absorption edges (11.865 keV) [10]

$$\left.\begin{aligned} A &= 2F_{0,1}F_{1,2}\left(1+F_{2,3}^2\right)/\left(1-F_{0,1}^2\right)\left(1-F_{1,2}^2\right)\left(1-F_{2,3}^2\right) \\ B &= 2F_{1,2}F_{2,3}\left(1+F_{0,1}^2\right)/\left(1-F_{0,1}^2\right)\left(1-F_{1,2}^2\right)\left(1-F_{2,3}^2\right) \\ C &= 2F_{0,1}F_{2,3}/\left(1-F_{0,1}^2\right)\left(1-F_{1,2}^2\right)\left(1-F_{2,3}^2\right) \\ D &= 2F_{0,1}F_{1,2}^2F_{2,3}/\left(1-F_{0,1}^2\right)\left(1-F_{1,2}^2\right)\left(1-F_{2,3}^2\right) \end{aligned}\right\}, \tag{9.16}$$

where $F_{j-1,j}$ are the Fresnel coefficients at the interfaces between the ($j$ - 1)th and $j$-th layers, and $t_j$ is the thickness of the $j$-th layer. Figure 9.11 shows enlarged drawing of the reflection profiles of Fig. 9.10 in the $k$ region above the critical wave vector. It is not too much to stress for the results of Fig. 9.11 that the entire oscillating structure is shifted to the lower $k$ side as the energy

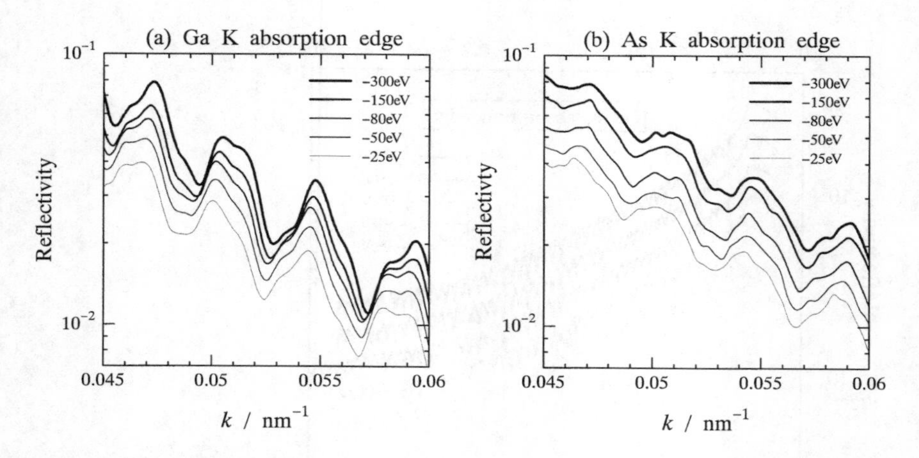


**Fig. 9.11. a, b** Enlarged drawings of the reflection profiles in the $k$ region above the critical wave vector at the lower-energy side of **a** the Ga-K and **b** the As-K absorption edges [10]

approaches an absorption edge from $-300$ to $-25$ eV in the $k$ region above the critical wave vector at both the Ga-K and As-K absorption edges. However, it is quite difficult to estimate quantitatively to what extent the critical wave vector of a layer may change, because the oscillating curves seem rather complicated owing to the mixture of several frequency components. Thus, it is essentially required to separate the oscillating part into each frequency component. For this purpose, it may be possible to use a least-squares curve-fitting technique to extract information such as the critical wave vector, the layer thickness and the interfacial roughness from the whole reflection curve. However, the actual implementation of this data processing is not so easy, because of the large number of parameters needed and their correlations, which could frequently lead to false minima in the fit and misinterpretation of the results. The author believes that the use of the Fourier filtering technique might be one way to overcome these problems.

Equations (9.14) and (9.15) indicate that the oscillating part is written as the sum of cosine functions in which $\gamma_j$ is the product of the film thickness, $t_j$, and the term $(k^2 - k_{cj}^2)^{1/2}$. Therefore, when the oscillating part is plotted as a function of $K \equiv (k^2 - k_{cj}^2)^{1/2}$, the distribution of thickness, $\Phi(t)$, can be obtained by Fourier transformation as follows [15]:

$$\Phi(t) = \int_0^{K_{\max}} K^n \chi(K, E) \exp(-4\pi \mathrm{i} K t) \mathrm{d}K. \tag{9.17}$$

For example, Fig. 9.12a gives the oscillating part, $K^3\chi(K, E)$, of the reflection profile at an energy of 25 eV below the Ga-K absorption edge obtained by subtracting the non-oscillating part. Here, the value of the critical wave

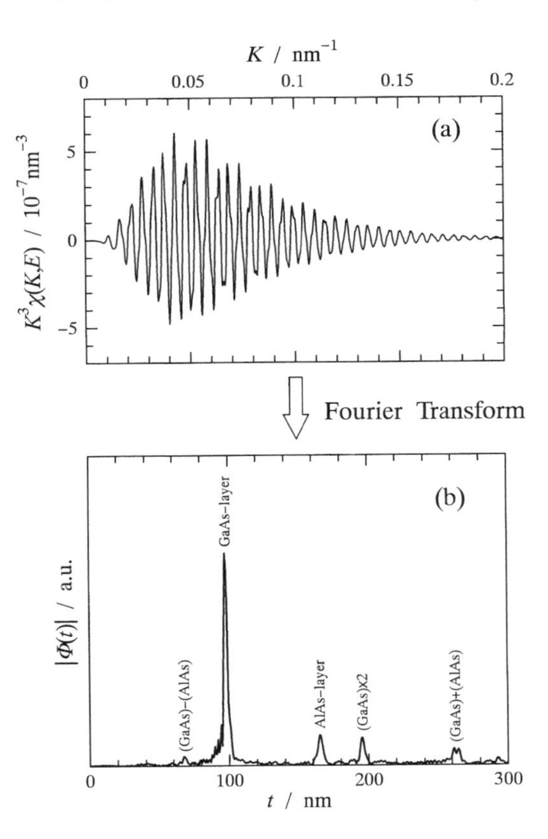


**Fig. 9.12.** **a** Oscillating part of $\chi(K,E)$ in the reflection profile of GaAs/AlAs/GaAs at an energy of 25 eV below the Ga-K absorption edge and **b** the magnitude of the Fourier transform [10]

vector ($k_c = 0.0335\ \mathrm{nm}^{-1}$ ) for the GaAs bulk is used. It may safely be said that a very accurate $k_{cj}$ value is not necessary at this stage as long as the resultant $\Phi(t)$ is not significantly distorted, because the present main purpose is not to determine the thickness of layers precisely. Figure 9.12b also shows the resultant magnitude of Fourier transformation, $|\Phi(t)|$, estimated from the oscillating data at the energy of 25 eV below the Ga-K absorption edge. In this figure, five peaks are clearly obtained. They qualitatively coincide with the thickness values of the GaAs layer (~100 nm), AlAs layer (~170 nm), their sum (~270 nm) and their difference (~70 nm). It is rather stressed here that a peak corresponding to a thickness twice as large as that of the GaAs layer can be also observed at about 200 nm, arising from the triple reflection in the GaAs layer. The difference between the thicknesses of the GaAs layer and the AlAs layer is sufficiently large that the oscillating profile, $K^3\chi_i(K,E)$, of an individual layer may be isolated in terms of the inverse

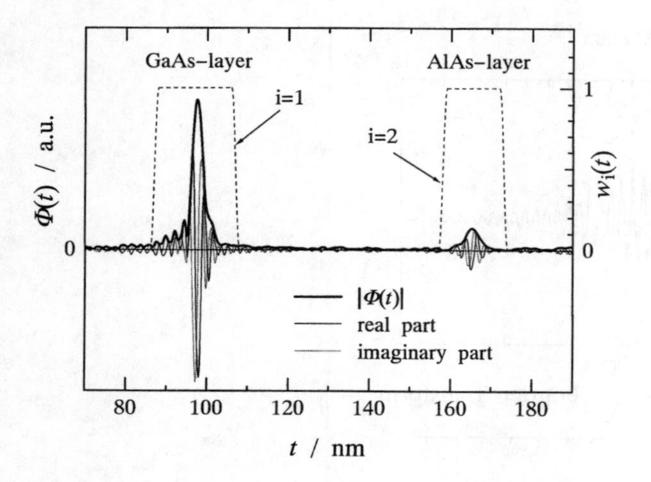


**Fig. 9.13.** Real and imaginary parts of thickness distribution. *Broken lines* denote the window functions presently used for selecting the thickness range for the GaAs layer and AlAs layer [10]

Fourier transform by selecting the thickness range of interest with a smooth window, $w_i(t)$, such as the Hanning function in the following equation [10]:

$$K^3\chi_i(K,E) = \int_{t_i^{\min}}^{t_i^{\max}} w_i(t)\Phi(t)\exp(4\pi\mathrm{i}Kt)\mathrm{d}t, \tag{9.18}$$

$$\left.\begin{aligned} w_i(t) &= \left[1-\cos\left\{\pi\left(t-t_i^{\min}\right)/d\right\}\right], && t_i^{\min} < t < t_i^{\max}+d \\ &= 1 && , t_i^{\min}+d < t < t_i^{\max}-d \\ &= \left[1-\cos\left\{\pi\left(t_i^{\max}-t\right)/d\right\}\right] && , t_i^{\max}-d < t < t_i^{\max} \end{aligned}\right\}, \tag{9.19}$$

where $d = 2$ nm is used in the present case. The window functions presently used for selecting the thickness range for the GaAs layer and AlAs layer are illustrated by broken lines in Fig. 9.13. The Fourier filtering may cause some distortions mainly in the amplitude. However, it is rather mentioning here that both the real and imaginary parts of $\Phi(t)$ are filtered identically with a sufficiently wide window so that phase error can be minimized. After filtering, the inverse Fourier transforms of (9.18) are calculated for obtaining the individual oscillating components, $\chi_1(K,E)$ for the GaAs layer and $\chi_2(K,E)$ for the AlAs layer. The results are shown in Fig. 9.14 together with the total oscillating profile, $\chi(K,E)$. The present results suggest that the Fourier filtering technique works rather well for separation into lower and higher frequency components.

Figure 9.15a shows the energy variation of $\chi_1$ and $\chi_2$ at the lower-energy side of the Ga-K absorption edge as a function of $K^2$. The energy variation

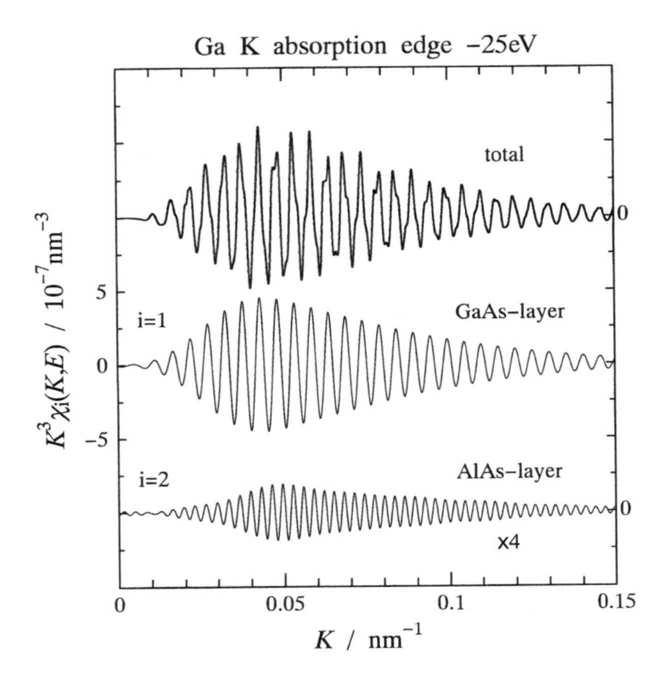


**Fig. 9.14.** Individual oscillating components, $\chi_1(K, E)$ for the GaAs layer and $\chi_2(K, E)$ for the AlAs layer, bound using the inverse Fourier transforms of (9.18). The total oscillating profile $\chi(K, E)$ is also given [10]

of $\chi_1$ and $\chi_2$ at the As-K absorption edge are given in Fig. 9.15b. It should be noted from the relation of (9.14) and (9.15) that the energy variation of phase in the oscillating component of $\chi_1$ and $\chi_2$ is exactly equal to the that in critical wave vector case given by $\Delta k_c^2(E)$ in Eq.(9.10), when plotted as a function of $K^2$. Systematic shift in phase of the oscillating component $\chi_1$ for the GaAs layer is clearly observed in Fig. 9.15a, whereas such phase variation is not detected in $\chi_2$ for the AlAs layer without Ga. On the other hand, appreciable phase shifts are also found in both $\chi_1$ and $\chi_2$ with the same magnitudes corresponding to the measurement at the lower-energy side of the As-K absorption edge (see Fig. 9.15b), and their magnitudes of shift also coincide with that for $\chi_1$ at the Ga-K absorption edge.

Figure 9.16 indicates plots of $\Delta k_c^2(E)$ estimated from the phase shifts against $f'_{\mathrm{Ga}}(E)$ or $f'_{\mathrm{As}}(E)$ for both the GaAs layer and the AlAs layer. A linear relationship between $\Delta k_c^2(E)$ and $f'(E)$ is clearly seen in each case. From each slope of the straight lines, the atomic number densities of Ga and As, $\rho_{\mathrm{Ga}}$ and $\rho_{\mathrm{As}}$, in the GaAs layer can be estimated to be $24 \pm 3\,\mathrm{nm}^{-3}$, where the uncertainty in these values corresponds to the probable error in the linear least-squares fitting. The value of $\rho_{\mathrm{As}}$ in the AlAs layer is also

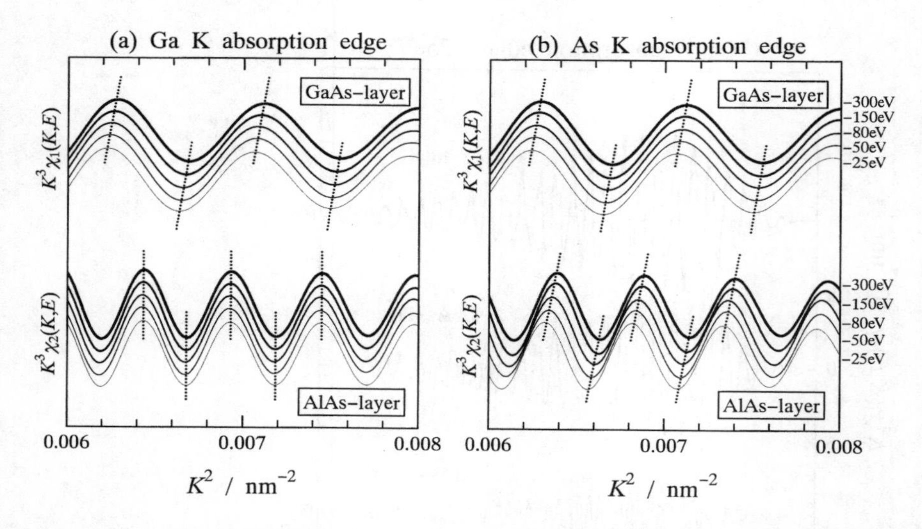


**Fig. 9.15.** Energy dependences of $\chi_1(K,E)$ and $\chi_2(K,E)$ for the GaAs layer and the AlAs layer at the lower-energy side of **a** the Ga-K absorption edge and **b** the As-K absorption edge [10]

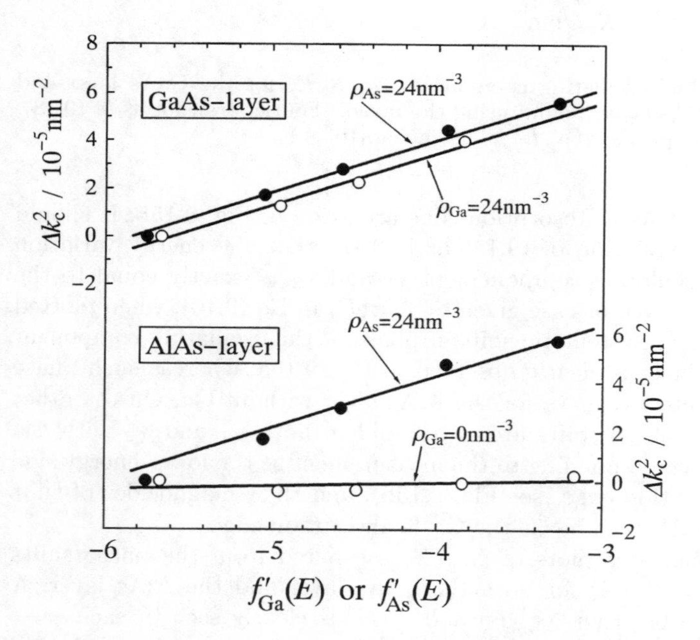


**Fig. 9.16.** Plots of $\Delta k_c^2(E)$ against $f'_{Ga}(E)$ (*open circles*) or $f'_{As}(E)$ (*solid circles*) for the GaAs layer and the AlAs layer [10]

estimated to be $24 \pm 3$ nm$^{-3}$, whereas the Ga element could not be detected in this layer as expected. These values are large, greater by about 10%, when compared to those of the bulk crystal structure ($\rho_{Ga}$ and $\rho_{As}$ are 22.1 nm$^{-3}$ for GaAs ($a_0$ = 0.5653 nm), and $\rho_{As}$ is 22.0 nm$^{-3}$ for AlAs ($a_0$ = 0.5661 nm)). Nevertheless, the agreement appears to be sufficient by considering many factors, including the experimental errors.

Analytical methods sensitive to the surface or the interface of materials are considered to be very desirable and important with regard to the requirements of recent progress both in materials science and advanced technology. The author takes the opinion that the capability of the AGXR method has been well confirmed for determining the atomic number density in multi-layered thin films, when coupled with the Fourier filtering technique. Even if the number of layers increases, each individual oscillating component could be isolated and analyzed separately by using the data processing given in this section, as long as a sufficient difference (for example, $\Delta t \geq 15$ nm) in the layer thickness holds in the Fourier transform. With respect to such a point, it would be very interesting to extend the AGXR method to the characterization of various multi-layered thin films including both crystalline and non-crystalline cases.

## References


1. D.H. Mosca, F. Petroff, A. Fert, P.A. Schreder, W.P. Pratt Jr., R. Laloee and S. Lequien: J.Magn. Mater., **94**, L1 (1991)
2. S.S. Parkin, R. Bhadra and K.P. Roche: Phys. Rev. Lett., **66**, 2152 (1991)
3. K.L. Weiner: Z.Krist., **123**, 315 (1966)
4. G.H. Vineyard: Phys. Rev., B **26**, 4146 (1982)
5. W.C. Marra, P. Eisenberger and A.Y. Cho: J. Appl. Phys., **50**, 6972 (1979)
6. J. Bohr, R. Feidenhans'l, M. Nielsen, M. Toney, R.L. Johnson and I.K. Robinson: Phys. Rev. Lett., **54**, 1275 (1985)
7. G. Lim, W. Parrish, C. Ortiz, M. Bellotto and M. Hart: J. Mater. Res., **2**, 471 (1987)
8. M. Saito, T. Kosaka, E. Matsubara and Y. Waseda: Mater. Trans., JIM, **36**, 1 (1995)
9. M. Saito, E. Matsubara and Y. Waseda: Mater. Trans. JIM, **37**, 39 (1996)
10. M. Saito and Y. Waseda: Mater. Trans. JIM, **40**, 1044 (1999)
11. R.W. James: *The Optical Principles of the Diffraction of X-rays* (G.Bells, London 1954)
12. M.F. Doerner and S. Brennan: J. Appl. Phys., **63**, 126 (1988)
13. L. Névot, P. Croce: Rev. Phys. Appl., **15**, 761 (1980)
14. Y. Sone, K. Yoshioka, M. Tochihara and O. Hashimoto: Kawasaki Steel Giho, **21**, 34 (1983)
15. K. Sakurai and A. Iida: *Advances in X-ray Analysis* (Plenum Press, New York 1992) Vol.35, pp. 813
16. M. Born and E. Wolf: *Principles of Optics 6th Edition* (Pergamon, New York 1980)

# 10. Merits of Anomalous X-ray Scattering and Its Future Prospects

There are many methods for structural investigation of various materials in a variety of states, and a large amount of experimental effort has been devoted to this field in the past. A number of techniques for X-rays, neutrons and others have been used to characterize the structure of both crystalline and non-crystalline materials, and each technique has, of course, its own advantages and disadvantages. The relative merits of these various techniques for structural characterization of materials have already been provided in detail in some specialized monographs (for example, [1,2,3,4]). The intention of this article is not to duplicate their description. We give here some essential points regarding the anomalous X-ray scattering (AXS) method with an emphasis on future developments.

## 10.1 Comparison of the AXS Analysis with the EXAFS Analysis

The extended X-ray absorption fine structure (EXAFS) signal, as shown in Fig. 10.1 using the results from an Ni film as an example [5], refers to the oscillatory modulation of the X-ray absorption coefficient within a few hundred eV beyond the absorption edge of a specific element in a system of interest. The EXAFS oscillations from the monotonic term in the absorption coefficient due to both K and L shells can now theoretically be interpreted by the effect arising from the interference of the outgoing photoejected electron with the backscattered one emerging from the near-neighbor surrounding atoms. Thus, the frequency of the EXAFS signal mainly depends on the correlation distance between the central absorbing atom and the neighboring atom on the one hand, and the amplitude of the EXAFS signal is strongly affected by the number and the backscattering ability of the neighboring atoms on the other. Hence, it is readily understood that information about neighboring atoms around a central absorbing atom, frequently referred to as EXAFS rdf, can be obtained from the EXAFS signal. It is also stressed that the EXAFS signal contains quantitatively accurate information about the local environmental structure for an absorbing atom, even though its content is relatively small. On the other hand, the radial distribution function (RDF) estimated

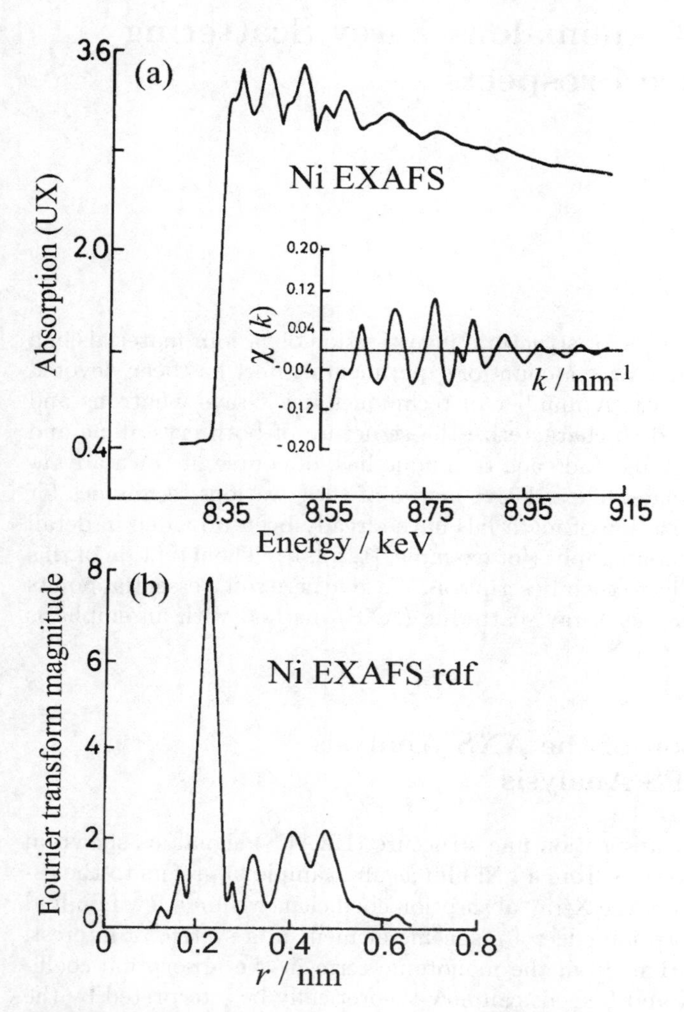


**Fig. 10.1.** Experimental spectrum of EXAFS for Ni foil with the fine structural function $\chi(k)$ [5]

from the usual X-ray diffraction data corresponds to the convolution of the EXAFS rdf, i.e. the average atomic distribution in a system as a function of radial distance. A more detailed discussion including data processing is available in specialized monographs on this relatively new technique (see, for example, [6, 7]).

The availability of a high-intensity X-ray source and the Fourier transformation analysis first proposed by Stern, Sayers and Lytle [8] have resulted in the establishment of the so-called EXAFS method as a tool for structural characterization of materials, although the EXAFS signal itself has been known for a long time. Thus, the EXAFS method is undoubtedly one of the most powerful methods, particularly for determining the local environmental structure around a specific element in both crystalline and non-crystalline materials. However, the AXS method is much more straightforward, at least, theoretically. For example, the theoretical difficulties associated with the electron phase shifts and mean free path still make it impossible to obtain reliable information from the EXAFS measurement alone, particularly for systems with unknown structures, such as liquids and glasses. This point was clearly stressed in 1981 by Lee et al. [9]. In a sense, the power of EXAFS may be somewhat overemphasized, although it has several impressive advantages. On the other hand, AXS does not require the information of phase shifts and mean free path.

The EXAFS measurement usually provides accurate information on the nearest-neighbor correlation for a specific atom. It is, however, relatively difficult to obtain information on the higher-order structure such as the second and third nearest neighbors, mainly due to the cut off (usually 30–50 $nm^{-1}$) in the low wave vector region [1], where the usual first peak appears in the interference function of non-crystalline materials. The AXS method reduces such difficulty by making available partial or environmental RDFs as a function of radial distance. This is particularly true in the discussion of structural features of metallic amorphous alloys produced by rapid quenching from the melt, because their structure is known to primarily be characterized by the second peak splitting in the RDF [10], as easily seen in the results of Fig. 10.2 [11,12].

The EXAFS signal corresponds to the energy dependence of the imaginary component, $f''$, of the anomalous dispersion factors. On the other hand, the intensity difference obtained from the particular measurement in AXS with two different energies where $f''$ is small and nearly constant provides information only about atoms scattering X-rays anomalously. In other words, this intensity difference should be attributed to the change due to only the real component, $f'$, of the specific atoms. For these reasons, the experimental data obtained by the energy-derivative technique in AXS (or differential anomalous scattering as named by Fuoss et al. [13]) is, in a sense, very similar to the information given by EXAFS, although there are differences in detail as mentioned previously. In order to facilitate an understanding of such a relation, a schematic for comparing X-ray diffraction with the EXAFS measurement is given in Fig. 10.3 using the case of non-crystalline materials [14].

It may be worth mentioning for non-crystalline materials that, AXS gives better coordination numbers, whereas EXAFS provides direct information regarding the nature of the atomic pair for the peaks. Therefore, AXS data could supplement EXAFS data or vice versa, and the EXAFS method is superior to AXS is in some special cases, such as in the study of the environmental structure of minor constituent elements in a sample.

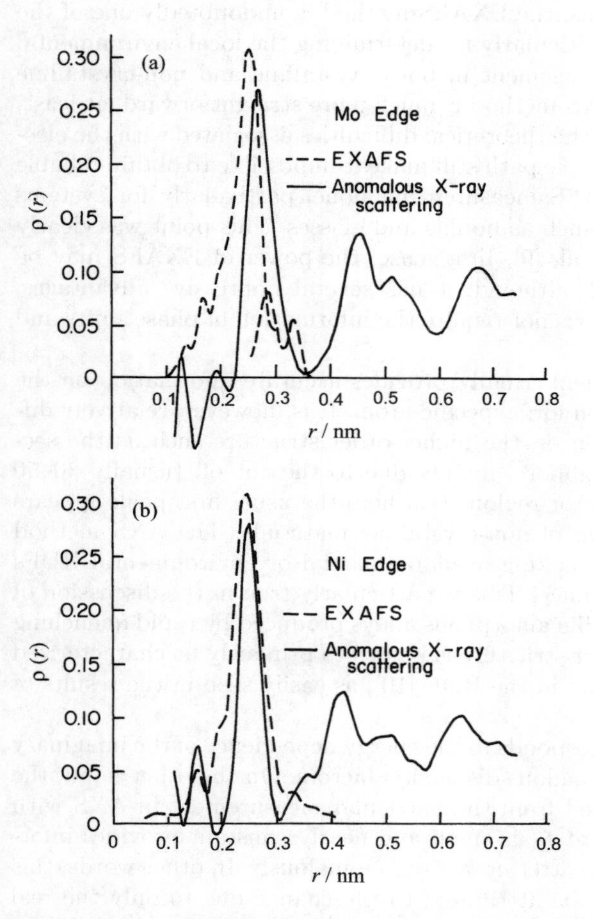


**Fig. 10.2. a, b** Environmental structural information **a** around Mo or **b** around Ni for the amorphous $Mo_{50}Ni_{50}$ alloy. *Solid lines*: AXS results [11]; *broken lines*: EXAFS results [12]

The AXS method could be applied to various systems with only a few exceptions, such as light elements, when coupled with the synchrotron-radiation source by making available a good continuous spectrum over a much wider energy range. This is an advantage in comparison with the AXS method and isotope substitution method with neutrons. Anomalous neutron scattering is applicable only to a small number of elements, such as $^{6}$Li, $^{10}$B, $^{113}$Cd, $^{149}$Sm, $^{153}$Eu and $^{157}$Gd, although the variation in the anomalous dispersion factors of neutrons is several times larger than in the X-ray case, as exemplified by Fig. 10.4 using the results of $^{113}$Cd [15]. The isotope substitution method is also limited by the number of appropriate isotopes.

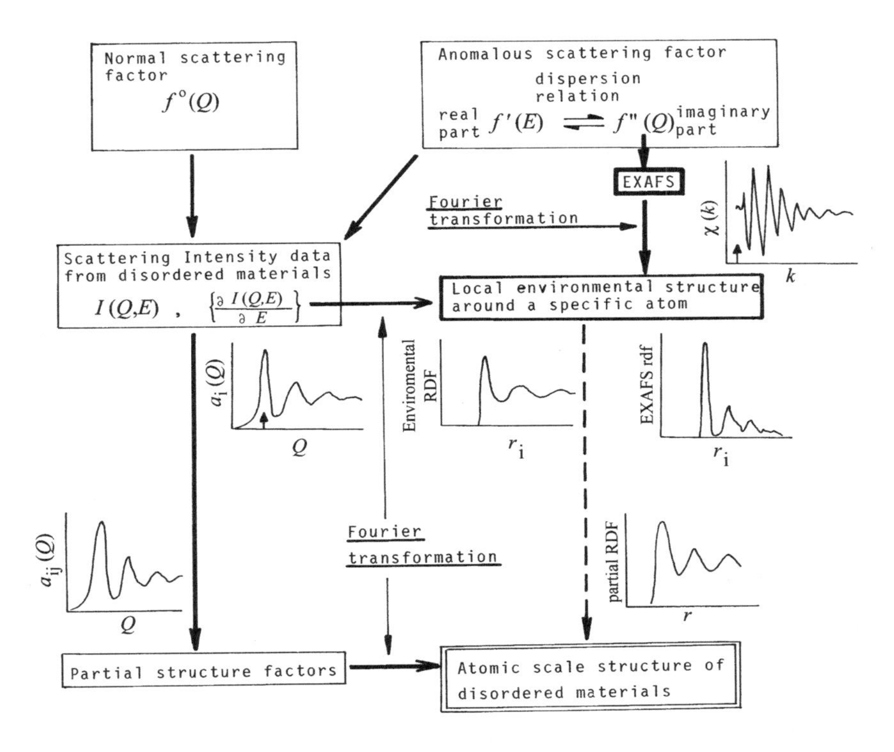


**Fig. 10.3.** Schematic of the relation between the AXS analysis and the EXAFS analysis for structural characterization of non-crystalline materials [14]

The structural investigation of solutions has been recognized as one of the most important research subjects in both aqueous solution chemistry and hydrometallurgy. Such importance is again emphasized in parallel with recent progress in modern biochemistry [2, 7, 16]. As shown in Chap. 7, the use of the AXS method will again bring about a significant breakthrough in this research subject by obtaining the average distribution of water molecules around a specific ion in solutions without complete separation of all partial structural functions. Similar information about hydration numbers in solution can be obtained from the first-order difference scattering of neutrons with isotopes (see, for example, [17]). Nevertheless, the AXS method may be superior to the neutron case where the structure is automatically assumed to remain identical upon substitution by the isotope. Such a structural assumption strongly depends on the systems and is frequently questionable, even if the isotopes are chemically similar and their size difference is small. For example, the low-angle behavior of water ($H_2O$) has been found to slightly differ from that of heavy water ($D_2O$), as suggested by Bosio et al. [18].

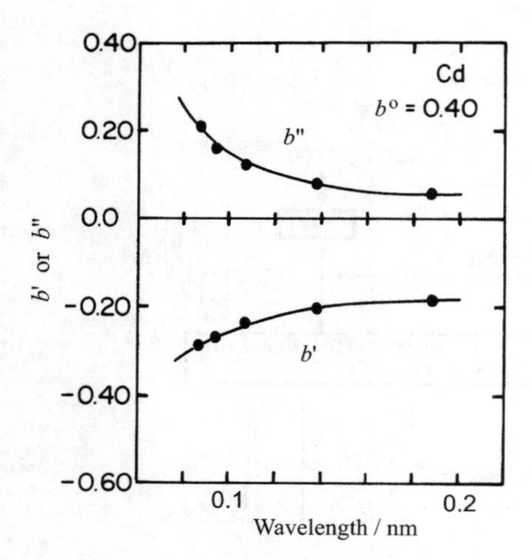


**Fig. 10.4.** Anomalous dispersion factors of $^{113}$Cd for neutrons [15]

The AXS method is, of course, not infalliable in the determination of the structure of materials. For example, unless a reasonably higher-energy absorption edge is used, the disadvantage due to the limited wave vector range available ($Q_{\max}$ values less than about 70 nm$^{-1}$) is unavoidable, and, in such a case, a careful interpretation of the resultant RDF is required. Nevertheless, as mentioned in this book with various selected examples, we have already built up a rather wider base for the AXS method, including the present status and future directions for its application to structural study of various materials. A coupled angular scanning mode using an energy-sensitive, solid-state detector and a synchrotron-radiation source unanimously appears to hold promise that the anomalous (resonance) X-ray scattering method should basically work very well, and its potential power, in author's view, may not be overemphasized. When this relatively new method is completed, it should be possible to have a significant impact on the relationships between atomic-scale structure and various properties of multi-component materials in a variety of states with much higher reliability.

In AXS for most of the elements the real part of the anomalous dispersion term, $f'$, is typically 15–20 % of the standard atomic scattering factor at the K-shell absorption edge, and $f'$ appears to have substantially larger value (over 30%) at the L-shell absorption edge, as shown in Fig. 10.5 [14]. This is based on the following reason: The intense X-ray source of synchrotron radiation provides not only high intensity but also a clean white spectrum over a much wider energy range above 4 keV, compared with the conventional white X-ray source, such as a W target. Only the energy above 15 keV is generally used for measurement with a conventional W target, because of the L lines of the W atoms and K$\alpha$ and K$\beta$ lines of the Mo atoms arising

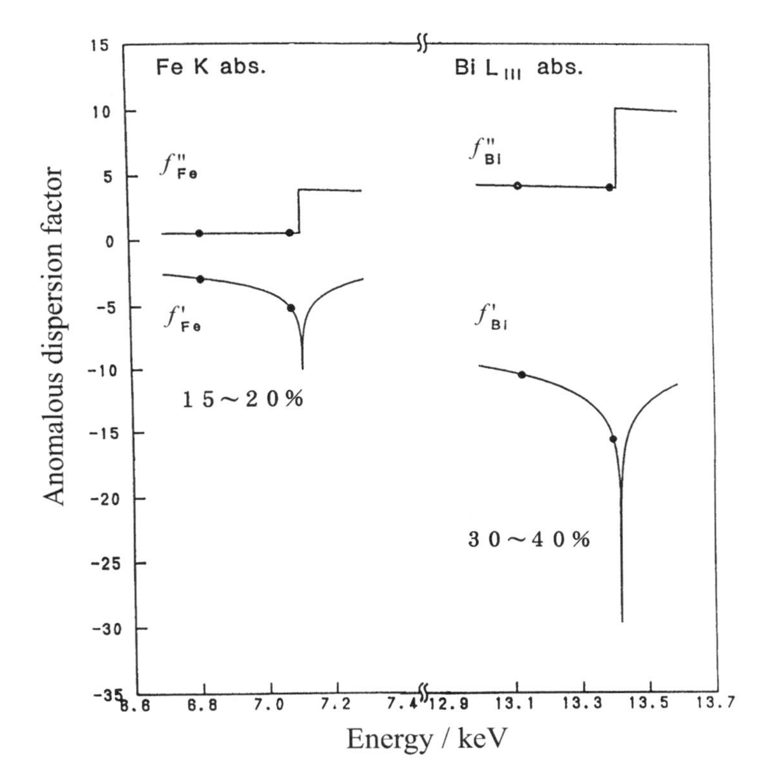


**Fig. 10.5.** Anomalous dispersion factors of Fe and Bi as a function of energy (calculated values)

from the impurities in the filament. Thus, this situation can improve both the acquisition and quality of the AXS effect. On the other hand, the synchrotron-radiation source is only limited by the absorption of the Be window.

If further investment with a highest priority can be continued in developing the use of the anomalous dispersion effect of X-rays with a synchrotron-radiation source, anomalous (resonance) X-ray scattering will achieve the goal in a relatively short period.

## 10.2 Energy Dependence (1 – 50 keV) of the X-ray Anomalous Dispersion Factors

The energy variation of X-ray anomalous dispersion factors provides a useful guide for selecting the experimental conditions for anomalous X-ray scattering (AXS) measurements. For this reason, X-ray anomalous dispersion factors, $f'$ and $f''$, for a total of 96 neutral atoms were estimated in the energy range between 1 and 50 keV; this information, including mass absorption coefficients is available as the public database SCM-AXS (http://www.iamp.tohoku.ac.jp/; http://www.tagen.tohoku.ac.jp/). The basic procedure of calculation in this work followed a method proposed by Cromer and Liberman [19], i.e. the relativistic estimation of $f'$ and $f''$ was made numerically without approximation to the form of the cross-section vs energy curve given in [20]. The present calculations were made with 48 points in the Gaussian integral at intervals of 1 eV in the close vicinity of the absorption edge. However, the following point may be suggested: The calculations at smaller intervals such as 0.1 eV indicate more distinct change in $f'$. Numerical examples are $-30.875$ for E = 5.0116 keV in comparison with $-22.519$ for E = 5.011 keV and $-24.266$ for E = 5.012 keV in the case of cesium.

Approach windows for this database are given in Fig. 10.6. For convenience in future investigations using the AXS method, the energies of absorption edges in keV units for various elements are summarized in Tables 10.1 and 10.2. The experimental information of the absorption edges for various elements is given in detail by Bearden [21]. It should also be kept in mind that the ionization state of an atom affects the position of the absorption edge. This is particularly true at energy levels close to the absorption edge, although such an energy shift is usually not so distinct.

This calculation was supported in part by the Ito Science Foundation and the Toshiba Corporate Research and Development Center (Kawasaki) through grants to the author.

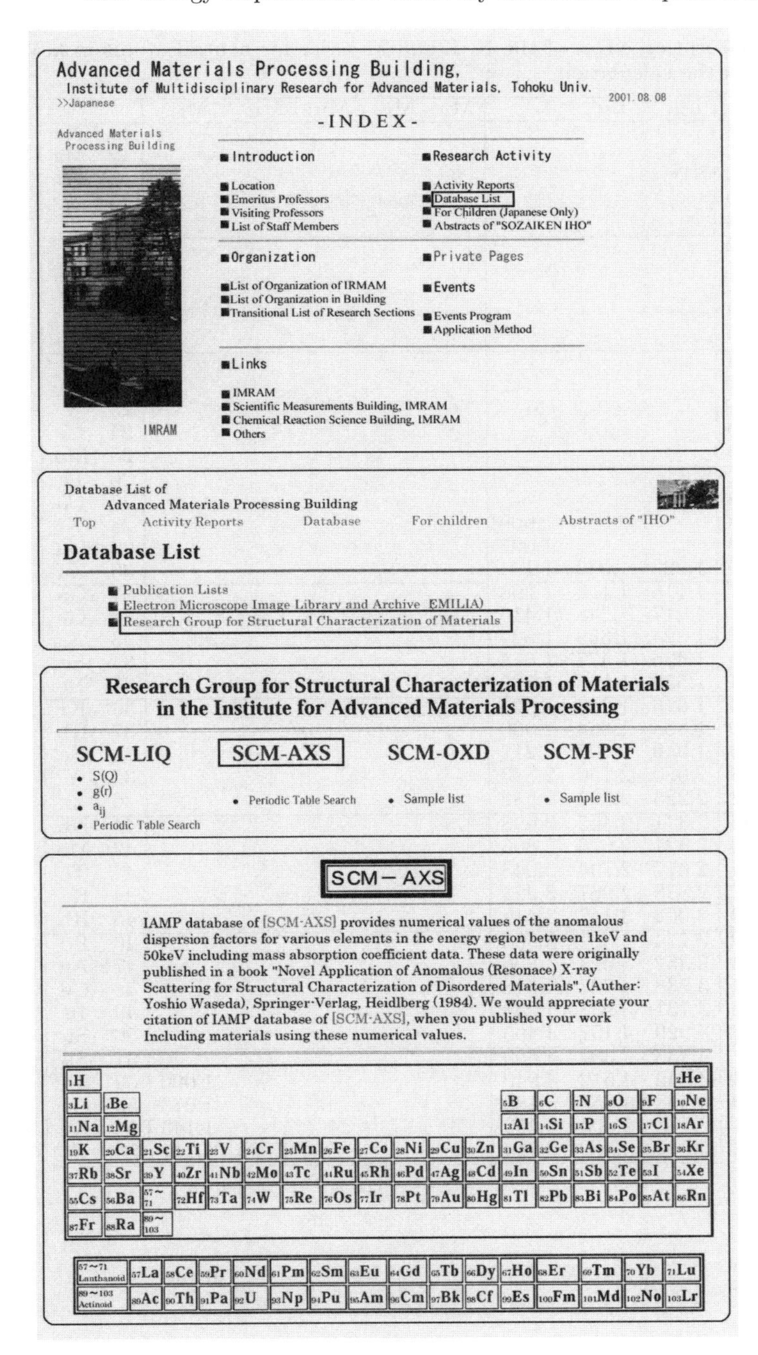


**Fig. 10.6.** Approach windows for the public database

**Table 10.1.** Absorption edges of the K, L and M series for neutral atoms in keV units obtained in the calculation

| | | K | $L_{III}$ | $L_{II}$ | $L_{I}$ | $M_{V}$ | $M_{IV}$ | $M_{III}$ | $M_{II}$ | $M_{I}$ | | |
|---|---|---|---|---|---|---|---|---|---|---|---|---|
| 11 | Na | 1.073 | | | | | | | | | 11 | Na |
| 12 | Mg | 1.305 | | | | | | | | | 12 | Mg |
| 13 | Al | 1.560 | | | | | | | | | 13 | Al |
| 14 | Si | 1.839 | | | | | | | | | 14 | Si |
| 15 | P | 2.146 | | | | | | | | | 15 | P |
| 16 | S | 2.472 | | | | | | | | | 16 | S |
| 17 | Cl | 1.823 | | | | | | | | | 17 | Cl |
| 18 | Ar | 3.203 | | | | | | | | | 18 | Ar |
| 19 | K | 3.608 | | | | | | | | | 19 | K |
| 20 | Ca | 4.039 | | | | | | | | | 20 | Ca |
| 21 | Sc | 4.493 | | | | | | | | | 21 | Sc |
| 22 | Ti | 4.967 | | | | | | | | | 22 | Ti |
| 23 | V | 5.466 | | | | | | | | | 23 | V |
| 24 | Cr | 5.990 | | | | | | | | | 24 | Cr |
| 25 | Mn | 6.539 | | | | | | | | | 25 | Mn |
| 26 | Fe | 7.112 | | | | | | | | | 26 | Fe |
| 27 | Co | 7.709 | | | | | | | | | 27 | Co |
| 28 | Ni | 8.333 | | | 1.009 | | | | | | 28 | Ni |
| 29 | Cu | 8.979 | | | 1.097 | | | | | | 29 | Cu |
| 30 | Zn | 9.659 | 1.020 | 1.043 | 1.194 | | | | | | 30 | Zn |
| 31 | Ga | 10.367 | 1.116 | 1.143 | 1.298 | | | | | | 31 | Ga |
| 32 | Ge | 11.103 | 1.217 | 1.248 | 1.415 | | | | | | 32 | Ge |
| 33 | As | 11.867 | 1.324 | 1.359 | 1.527 | | | | | | 33 | As |
| 34 | Se | 12.658 | 1.436 | 1.477 | 1.654 | | | | | | 34 | Se |
| 35 | Br | 13.474 | 1.550 | 1.596 | 1.782 | | | | | | 35 | Br |
| 36 | Kr | 14.326 | 1.675 | 1.728 | 1.921 | | | | | | 36 | Kr |
| 37 | Rb | 15.200 | 1.805 | 1.864 | 2.066 | | | | | | 37 | Rb |
| 38 | Sr | 16.105 | 1.940 | 2.007 | 2.217 | | | | | | 38 | Sr |
| 39 | Y | 17.038 | 2.080 | 2.156 | 2.373 | | | | | | 39 | Y |
| 40 | Zr | 17.998 | 2.223 | 2.307 | 2.532 | | | | | | 40 | Zr |
| 41 | Nb | 18.986 | 2.371 | 2.465 | 2.698 | | | | | | 41 | Nb |
| 42 | Mo | 20.000 | 2.521 | 2.626 | 2.866 | | | | | | 42 | Mo |
| 43 | Tc | 21.044 | 2.677 | 2.794 | 3.043 | | | | | | 43 | Tc |
| 44 | Ru | 22.117 | 2.838 | 2.967 | 3.224 | | | | | | 44 | Ru |
| 45 | Rh | 23.220 | 3.004 | 3.147 | 3.412 | | | | | | 45 | Rh |
| 46 | Pd | 24.350 | 3.174 | 3.331 | 3.605 | | | | | | 46 | Pd |
| 47 | Ag | 25.514 | 3.352 | 3.524 | 3.806 | | | | | | 47 | Ag |
| 48 | Cd | 26.711 | 3.538 | 3.727 | 4.018 | | | | | | 48 | Cd |
| 49 | In | 27.940 | 3.731 | 3.938 | 4.238 | | | | | | 49 | In |
| 50 | Sn | 29.200 | 3.929 | 4.157 | 4.465 | | | | | | 50 | Sn |
| 51 | Sb | 30.491 | 4.133 | 4.381 | 4.699 | | | | | | 51 | Sb |
| 52 | Te | 31.814 | 4.342 | 4.612 | 4.940 | | | | | 1.006 | 52 | Te |
| 53 | I | 33.169 | 4.558 | 4.853 | 5.189 | | | | | 1.073 | 53 | I |
| 54 | Xe | 34.561 | 4.783 | 5.104 | 5.453 | | | | | 1.145 | 54 | Xe |
| 55 | Cs | 35.982 | 5.012 | 5.360 | 5.714 | | | | 1.065 | 1.217 | 55 | Cs |

**Table 10.2.** Absorption edges of the K, L and M series for neutral atoms in keV units obtained in the calculation

| | K | $L_{III}$ | $L_{II}$ | $L_I$ | $M_V$ | $M_{IV}$ | $M_{III}$ | $M_{II}$ | $M_I$ | |
|---|---|---|---|---|---|---|---|---|---|---|
| 56 Ba | 37.438 | 5.247 | 5.624 | 5.989 | | | 1.063 | 1.137 | 1.293 | 56 Ba |
| 57 La | 38.922 | 5.483 | 5.891 | 6.262 | | | 1.124 | 1.205 | 1.362 | 57 La |
| 58 Ce | 40.441 | 5.724 | 6.164 | 6.549 | | | 1.186 | 1.273 | 1.435 | 58 Ce |
| 59 Pr | 41.988 | 5.964 | 6.440 | 6.835 | | | 1.243 | 1.338 | 1.511 | 59 Pr |
| 60 Nd | 43.566 | 6.208 | 6.722 | 7.120 | | | 1.298 | 1.403 | 1.576 | 60 Nd |
| 61 Pm | 45.189 | 6.459 | 7.013 | 7.428 | 1.027 | 1.052 | 1.357 | 1.472 | 1.647 | 61 Pm |
| 62 Sm | 46.831 | 6.716 | 7.312 | 7.737 | 1.090 | 1.106 | 1.420 | 1.541 | 1.723 | 62 Sm |
| 63 Eu | 48.516 | 6.977 | 7.617 | 8.052 | 1.131 | 1.161 | 1.481 | 1.614 | 1.800 | 63 Eu |
| 64 Gd | 50.236 | 7.243 | 7.930 | 8.376 | 1.186 | 1.218 | 1.544 | 1.688 | 1.881 | 64 Gd |
| 65 Tb | 51.993 | 7.514 | 8.252 | 8.708 | 1.242 | 1.275 | 1.612 | 1.768 | 1.968 | 65 Tb |
| 66 Dy | 53.785 | 7.790 | 8.581 | 9.046 | 1.295 | 1.333 | 1.676 | 1.842 | 2.047 | 66 Dy |
| 67 Ho | 55.615 | 8.071 | 8.918 | 9.394 | 1.352 | 1.392 | 1.742 | 1.923 | 2.129 | 67 Ho |
| 68 Er | 57.482 | 8.358 | 9.264 | 9.751 | 1.410 | 1.454 | 1.812 | 2.006 | 2.207 | 68 Er |
| 69 Tm | 59.386 | 8.648 | 9.617 | 10.115 | 1.468 | 1.515 | 1.885 | 2.090 | 2.307 | 69 Tm |
| 70 Yb | 61.329 | 8.944 | 9.978 | 10.486 | 1.528 | 1.577 | 1.950 | 2.173 | 2.398 | 70 Yb |
| 71 Lu | 63.310 | 9.244 | 10.348 | 10.870 | 1.589 | 1.640 | 2.024 | 2.264 | 2.492 | 71 Lu |
| 72 Hf | 65.347 | 9.561 | 10.739 | 11.270 | 1.662 | 1.717 | 2.108 | 2.366 | 2.601 | 72 Hf |
| 73 Ta | 67.412 | 9.881 | 11.135 | 11.681 | 1.735 | 1.794 | 2.194 | 2.469 | 2.708 | 73 Ta |
| 74 W | 69.521 | 10.206 | 11.543 | 12.099 | 1.810 | 1.872 | 2.281 | 2.575 | 2.820 | 74 W |
| 75 Re | 71.672 | 10.535 | 11.958 | 12.526 | 1.883 | 1.949 | 2.368 | 2.682 | 2.932 | 75 Re |
| 76 Os | 73.866 | 10.870 | 12.384 | 12.967 | 1.960 | 2.031 | 2.458 | 2.792 | 3.049 | 76 Os |
| 77 Ir | 76.107 | 11.215 | 12.823 | 13.418 | 2.041 | 2.116 | 2.551 | 2.909 | 3.174 | 77 Ir |
| 78 Pt | 78.390 | 11.563 | 13.272 | 13.879 | 2.122 | 2.202 | 2.646 | 3.027 | 3.296 | 78 Pt |
| 79 Au | 80.720 | 11.918 | 13.733 | 14.351 | 2.206 | 2.291 | 2.743 | 3.148 | 3.425 | 79 Au |
| 80 Hg | 83.097 | 12.283 | 14.208 | 14.838 | 2.295 | 2.385 | 2.847 | 3.279 | 3.562 | 80 Hg |
| 81 Tl | 85.525 | 12.657 | 14.697 | 15.346 | 2.390 | 2.485 | 2.957 | 3.416 | 3.704 | 81 Tl |
| 82 Pb | 87.999 | 13.034 | 15.119 | 15.860 | 2.484 | 2.586 | 3.067 | 3.554 | 3.851 | 82 Pb |
| 83 Bi | 90.521 | 13.418 | 15.710 | 16.387 | 2.580 | 2.688 | 3.177 | 3.697 | 3.999 | 83 Bi |
| 84 Po | 93.100 | 13.813 | 16.243 | 16.938 | 2.683 | 2.798 | 3.302 | 3.854 | 4.150 | 84 Po |
| 85 At | 95.724 | 14.213 | 16.784 | 17.492 | 2.787 | 2.909 | 3.426 | 4.008 | 4.317 | 85 At |
| 86 Rn | 98.398 | 14.619 | 17.336 | 18.048 | 2.893 | 3.022 | 3.538 | 4.159 | 4.482 | 86 Rn |
| 87 Fr | 101.130 | 15.030 | 17.905 | 18.638 | 3.000 | 3.136 | 3.663 | 4.327 | 4.652 | 87 Fr |
| 88 Ra | 103.920 | 15.443 | 18.483 | 19.236 | 3.105 | 3.249 | 3.792 | 4.490 | 4.822 | 88 Ra |
| 89 Ac | 106.750 | 15.870 | 19.082 | 19.839 | 3.219 | 3.370 | 3.909 | 4.656 | 5.002 | 89 Ac |
| 90 Th | 109.640 | 16.299 | 19.692 | 20.471 | 3.332 | 3.491 | 4.046 | 4.832 | 5.182 | 90 Th |
| 91 Pa | 112.590 | 16.732 | 20.313 | 21.103 | 3.442 | 3.611 | 4.174 | 5.001 | 5.367 | 91 Pa |
| 92 U | 115.600 | 17.165 | 20.946 | 21.756 | 3.552 | 3.728 | 4.304 | 5.182 | 5.548 | 92 U |
| 93 Np | 118.670 | 17.609 | 21.599 | 22.425 | 3.666 | 3.851 | 4.435 | 5.366 | 5.723 | 93 Np |
| 94 Pu | 121.810 | 18.056 | 22.265 | 23.096 | 3.708 | 3.973 | 4.557 | 5.541 | 5.933 | 94 Pu |
| 95 Am | 125.020 | 18.503 | 22.943 | 23.772 | 3.887 | 4.092 | 4.667 | 5.710 | 6.121 | 95 Am |
| 96 Cm | 128.210 | 18.929 | 23.778 | 24.459 | 3.971 | 4.227 | 4.797 | 5.895 | 6.288 | 96 Cm |
| 97 Bk | 131.580 | 19.451 | 24.384 | 25.274 | 4.132 | 4.366 | 4.977 | 6.147 | 6.556 | 97 Bk |
| 98 Cf | 135.950 | 19.929 | 25.249 | 26.108 | 4.253 | 4.497 | 5.109 | 6.359 | 6.754 | 98 Cf |

## References


1. B.K. Teo: *EXFAS Basic Principles and Data Analysis* (Springer, Berlin, Heidelberg, New York 1986)
2. G. Metrlik, C.J. Sparks and K. Fischer (editors): *Resonant Anomalous X-ray Scattering* (North-Holland Amsterdam 1994)
3. D.K. Bowen and B.K. Tanner: *High Resolution X-ray Diffractometry and Topography* (Taylor & Francis, London 1998)
4. J. Daillant and A. Gibaud: *X-ray and Neutron Reflectivity: Principles and Applications* (Springer, Berlin, Heidelberg, New York 1999)
5. H. Oyanagi and S. Hosoya: J. Crystallogr. Soc. Jpn., **22**, 57 (1980)
6. P.A. Lee and B.K. Teo (Editors): *EXAFS* (Spectroscopy, Plenum Press, New York 1981)
7. D.C. Koningsberger and R. Prins: *X-ray Absorption; Principles, Applications, Techniques of EXAFS, SEXAF and XANES* (Wiley, New York 1988)
8. E.A. Stern, D.E. Sayers and F.W. Lytle: Phys. Rev., **11**, 4836 (1975)
9. P.A. Lee, P.H. Citrin, P. Eisenberger and B.M. Kincaid: Rev. Mod. Phys., **53**, 761 (1981)
10. Y. Waseda: *The Structure of Non-Crystalline Materials* (McGraw-Hill, New York 1980)
11. S. Aur, D. Kofalt, Y. Waseda, T. Egami, R. Wang, H.S. Chen and B.K. Teo: Solid State Commun., **48**, 111 (1984)
12. B.K. Teo, H.S. Chen, R. Wang and M.R. Antonio: J. Non-Cryst. Solids, **58**, 249 (1983)
13. P.H. Fuoss, P. Eisengerger, W.K. Warburton and A. Bienenstock: Phys. Rev. Lett., **46**, 1537 (1981)
14. Y. Waseda: *Novel Application of Anomalous X-ray Scattering for Structural Characterization of Disordered Materials* (Springer, Berlin, Heidelberg, New York 1984)
15. S.W. Peterson and H.G. Smith: Phys. Rev. Lett., **6**, 7 (19961)
16. J.K. Blasie and J. Stamatoff: Ann. Rev. Biophys. Bioeng., **40**, 451 (1981)
17. J.E. Enderby and G. Neilson: Rep. Prog. Phys., **44**, 593 (1981)
18. L. Boiso, J. Texeira and H.E. Stanley: Phys. Rev. Lett., **46**, 597 (1981)
19. D.T. Cromer and D. Liberman: J.Chem. Phys., **53**, 1891 (1970)
20. D.T. Cromer and D. Liberman: Los Alamos Scientific Laboratory Report, No. LA-4403 (LANL, Los Alamos 1970)
21. J.A. Bearden: Rev. Mod. Phys. **39**, 78 (1967)

# Index

Printing: Mercedes-Druck, Berlin
Binding: Stein+Lehmann, Berlin